Electric Infrastructure Protection (EPRO®)

Handbook V

"Blackstart Power Restoration for a Greener Grid"

Black Sky Resilience for a Changing Electric Grid Best Practices Handbook

EIS COUNCIL

EPRO Handbook V

"Blackstart Power Restoration for a Greener Grid"

Black Sky Resilience for a Changing Electric Grid Best Practices Handbook

First Edition – April 15, 2024
First Printing – September 1, 2024

ISBN: 978-0-9971659-5-1

Principal Author

Dr. Paul Stockton; Former U.S. Assistant Secretary of Defense for Homeland Defense

Co-author

Dr. Chris Beck; Chief Scientist and Vice President for Policy, EIS Council

Contributing Author and Researcher

Michael Ross; Mechanical & Aerospace Engineering, Syracuse University

Contributing Editors

Avi Schnurr; CEO and President, EIS Council

David Roop; Former Director, Electric Transmission, Dominion Energy; National Academy of Engineering Member; Past President, CIGRE U.S. National Committee

Steven Naumann; Former Vice President, Transmission and NERC Policy, Exelon Corporation

Tom Galloway; President and CEO, North American Transmission Forum

Frank Koza; Electric Subsector Coordinator, EIS Council

Clayton Clem; Former Vice President, Tennessee Valley Authority

Additional Peer Reviewers

Andrew Balascak, NATF
Gladys Brown Dutrieuille, NARUC
Kevin Curtis, Quanta Technologies
James Ellis, VLE Enterprises LLC
Lynna Estep, NATF
Nomi Etzion, NOGA Israel
Celeste Fay, Thorndike Mill
William Fay, Thorndike Mill
Dr. Steven Flynn, Northeastern University
Susan Ginsberg, Criticality Sciences
Wendy Goldsmith, Sustainability Visions
Ian Grant, TVA
Sherrell Greene, Advanced Technology Insights, LLC
Lordez Gutierrez, FEMA
Mary-Anna Holden, NJBPU (emeritus)
Curtis Ivins, PCS
Ronald Johnson, Thorndike Mill
Mark Jones, DHS, CISA
Arthur Mouco, Criticality Sciences
Ron Mathai, TVA
Richard Mroz, NARUC
Darlene Phillips, PJM
Chris Pilong, PJM
Wanda Reder, GridX Partners
Jim Robb, NERC
Benjamin L. Ruddell, Criticality Sciences
Timothy Schwartz, PEPCO
Gary Soto-Fernandez, LUMA
John Twitchell, EIS Council
Lee Underwood, NATF
Justin Verbarendse, DHS, CISA
Christopher Weigand, EPRI
Paul Williams, EIS Council
Michael Zhang, PJM

ACKNOWLEDGEMENTS

Development and production of this Handbook relied on insights, consultations, and review from many individuals and organizations. Although they cannot all be acknowledged here, the editors and publisher wish to express their deepest appreciation to the many contributors to this Handbook, worldwide.

Foundations and Philanthropists

The project would also have been impossible without the generous support and active encouragement of philanthropic foundations and individuals. The authors and publisher wish to offer special commendation and thanks to these dedicated, visionary philanthropists and organizations.

Anonymous Donors

Published by

Electric Infrastructure Security (EIS) Council

Image Copyright Disclaimer

All images included in the Handbook are entirely the work of their respective authors, and their use does not imply the endorsement of EIS Council, the Handbook, or any of its related products by the authors. Images containing U.S. Government personnel or property do not imply the endorsement of EIS Council by the United States Government.

All images are unaltered, and free to use either as part of the Public Domain, through purchase of publication authorization, or with appropriate attribution and credit given to the respective authors.

U.S. Department of Defense Review Statement

This Handbook was reviewed for National Security purposes and cleared for publication by the U.S. Department of Defense.

The views expressed in this publication are those of the author and do not necessarily reflect the official policy or position of the Department of Defense or the U.S. government.

The public release clearance of this publication by the Department of Defense does not imply Department of Defense endorsement or factual accuracy of the material.

Table of Contents

EXECUTIVE SUMMARY 13

INTRODUCTION 15

Key findings and options for consensus-building on blackstart initiatives..18

A. Finding: Societal Continuity is Increasingly Dependent On Resilient Blackstart Capabilities 18

B. Finding: Blackstart Resilience Planning Must Include All Phases of the Restoration Process..21

C. Finding: Grid Operators Should Be Ready to Conduct Blackstart in the Midst of Societal Chaos and Intense Political Pressures 26

D. Finding: Dangers to Blackstart Resource Adequacy Are Growing.......29

E. Finding: Immediate Initiatives Are Required to Ensure the Operating Reliability of Blackstart in an IBR-Heavy Grid 33

F. Over the Longer Term, Ibrs Could Offer Unique Benefits to Blackstart Resilience – But Only If Their Technical and Operational Problems Can Be Resolved 35

G. Finding: Grid Following and Grid Forming Inverters Pose Distinct Problems – and Different Potential Advantages – for Blackstart Resilience 38

H. Finding: Measures to Secure the Flow of Fuel for Gas-Fired Generators Must Scale With Intensifying Threats................................40

I. Finding: New Hydropower, Geothermal and Fuel Cell Initiatives Can Help Meet Emerging Blackstart Requirements.............................44

J. Finding: Existing Nuclear Power Plants Present Critical Black Sky Challenges, But Could Become Future Black Sky Blackstart Assets ...46

CHAPTER ONE - BLACKSTART IN A TRANSFORMING GRID............49

INTRODUCTION: THE NATURE OF THE CHALLENGE........................51

I | BLACKSTART 101 59

Background 59

Initial Steps in the Blackstart Process 62

II | FERC, NERC, AND CURRENT REGULATORY FOUNDATIONS OF BLACKSTART RESILIENCE IN NORTH AMERICA 74

Blackstart Standards 75

III | BUILDING ON THE REGULATORY FOUNDATIONS OF BLACKSTART: STRUCTURE OF THE HANDBOOK 78
A. Training, Simulating and Exercising Blackstart for Real-World Blackstart Conditions 79
B. Operating in Blackout Conditions 83
C. The DARPA RADICS Program as a Best Practice 85
D. Modeling and Simulation 88
E. Providing Security for Restoration Personnel 88
STRUCTURE OF THE HANDBOOK 89

CHAPTER TWO - THE IMPORTANCE OF BLACKSTART FOR SOCIETAL CONTINUITY AND NATIONAL SECURITY 91

INTRODUCTION 93

I | BLACKSTART TO SUPPORT SOCIETAL CONTINUITY 95
A. Community Lifelines 96
B. Electrification: Implications for Blackstart and the Prioritized Restoration of Power 99

II | ARTIFICIAL INTELLIGENCE AS ENABLER OF BLACKSTART RESTORATION – AND AN INTENSIFYING THREAT 105

III | EMERGENCY POWER AND ADVANCED MICROGRIDS FOR NCFs AND COMMUNITY LIFELINES 110
A. Emergency Power Generation 111
B. Advanced Microgrids 115

IV | BLACKSTART: A BULWARK FOR NATIONAL SECURITY IN THE UNITED STATES AND ITS GLOBAL PARTNERS 118
A. Blackstart Preparedness against Chinese Attacks 119
B. Employing Emergency Authorities to Shape Blackstart Restoration 122

CHAPTER THREE - ENSURING RESOURCE ADEQUACY FOR BLACKSTART IN A DECARBONIZING GRID 127

INTRODUCTION 129

I | THE RELIABILITY CONTEXT FOR BLACKSTART 138

II | RESOURCE ADEQUACY 143
A. The Changing Generation Fleet 144
B. Blackstart Resource Adequacy 149

CHAPTER FOUR - OPERATING RELIABILITY FOR RESTORATION SYSTEMS IN AN IBR-HEAVY GRID.. 165

INTRODUCTION...167

I | IBR PERFORMANCE DEFICIENCIES: IMPLICATIONS FOR BPS RELIABILITY AND POWER RESTORATION..........................168

A. Resilience for IBRs – Including for Blackstart..172

B. Blackstart in a Hybrid Electric System..174

II | ESSENTIAL RELIABILITY SERVICES FOR INCIDENT RESPONSE AND BLACKSTART ISLAND STABILITY....................177

A. Ramping as an Essential Reliability Service 178

B. Frequency Response, Voltage Support, and Advanced IBR Capabilities 182

CHAPTER FIVE - ENSURING RESILIENT NATURAL GAS SUPPLIES FOR BLACKSTART RESTORATION... 187

INTRODUCTION...189

I | THE STRATEGIC SUPPORT FORCE OF THE PEOPLE'S LIBERATION ARMY 191

A. Implications of the Colonial Pipeline Attack for Blackstart Fuel Resilience 193

II | SPECIFIC ATTACK VECTORS...196

III | COLLABORATIVE MEASURES TO STRENGTHEN THE RESILIENCE OF BLACKSTART FUEL SUPPLIES...........................199

IV | SECONDARY FUELS AS A SPECIAL FOCUS FOR ACTION..........205

V | REGULATORY AND COST RECOVERY INITIATIVES IN A DECARBONIZING GRID 208

VI | MANDATORY RELIABILITY AND CRITICAL INFRASTRUCTURE PROTECTION STANDARDS FOR NATURAL GAS SYSTEMS..212

CHAPTER SIX - EMPLOYING INVERTER-BASED RESOURCES FOR BLACKSTART RESTORATION......215

INTRODUCTION......217

Gas Disruptions, IBRs, and Broader Implications for Blackstart Resilience 219

I | IBRS FOR BLACKSTART – POTENTIAL RESILIENCE BENEFITS 223

II | MULTI-ISLAND PLANNING AND OPERATIONS......226

III | BLACKSTART COST RECOVERY, REGULATIONS, AND SECURITY IN A CHANGING RESOURCE MIX 229

B. Blackstart Resources: Security Versus Availability......230

C. Financial Incentives......234

D. Employing DERs for Blackstart Versus BES-Connected Assets......237

IV | EMPLOYING GRID-FOLLOWING AND GRID-FORMING IBRs FOR BLACKSTART......239

A. Grid-Following vs Grid Forming Inverters......241

B. Specifications for GFM Inverters: Making Survivability a New Priority247

V | DECENTRALIZATION AND GENERATION DIVERSITY: IBR-ENABLED STRATEGIES FOR BLACKSTART RESILIENCE...250

CHAPTER SEVEN - ADDITIONAL GENERATION OPTIONS FOR BLACKSTART OPERATIONS 253

I | SMALL HYDROPOWER GENERATION FOR BLACKSTART AND SYSTEM RESILIENCE 255

II | OTHER POSSIBLE BLACKSTART CRANKING PATHS SCENARIOS: BEYOND BES-LEVEL GENERATION 260

III | GEOTHERMAL GENERATION FOR BLACKSTART AND SYSTEM RESILIENCE 264

IV | NUCLEAR POWER: FLEXIBLE OPERATIONS AND SMALL MODUALR REACTORS (SMR)......269

A. Motivation and Overview......269

B. Flexible Operations as an Enabler for Extended Safe Shutdown and/or Load Following......271

C. Small Modular Reactors (SMRs) 272

D. Current, Promising Small Modular and Advanced Reactor Designs, ..274

E. Blackstart Capability of Small Modular Reactors......281

V | FUEL CELLS AS BLACKSTART ASSETS 283

CHAPTER EIGHT - MEETING RESILIENCE REQUIREMENTS AGAINST BLACK SKY HAZARDS 287

I | PLANNING 289
- A. Minimum Grid Concept 289
- B. Strategies for Load Pickup and Prioritization, Especially for Societal Continuity 291
- C. Minimum Grid Planning Steps 292

II | ASSET HARDENING 295
- A. Cyber Protection 296
- B. Earthquake Protection for Natural Gas Pipelines 298
- C. Electromagnetic Pulse (EMP) and Geomagnetic Disturbances (GMD) Protection 300
- D. Extreme Weather (Wind, Rain, Snow, Ice, Fire, Flood, Extreme Temperatures) Mitigation 305
- E. Kinetic Attack (Rifle fire, Explosives) Protection 306

III | SPARING / REDUNDANCY / STRATEGIC ISLANDING 308

IV | PERSONNEL 311
- A. Training 311
- B. Redundancy and Availability Measures 312

V | ESSENTIAL, WIDELY DEPLOYED BLACK SKY-CLASS RESILIENCE CAPABILITIES 313
- A. Emergency Communications 313
- B. Chaos Management 317
- C. Long Duration Power Cells 319

VI | HUMAN CONTINUITY 326
- A. Background 326
- B. The wider problem: systemic, global infrastructure fragility and expanding threats 327
- C. The Opportunity: Investor-driven, collaborative global resilience 328
- D. The Plan 330

CHAPTER NINE - CONCLUSION: FINDINGS, RECOMMENDATIONS, AND THE PATH AHEAD 333

The Path Ahead 341

The ELECTRIC INFRASTRUCTURE PROTECTION (EPRO®) HANDBOOK SERIES

A source of peer-reviewed analysis, recommendations and practical, implementable best practices to help infrastructure owners and operators, government agencies, and non-governmental organizations bolster resilience against Black Sky hazards.

EXECUTIVE SUMMARY

INTRODUCTION

Never before has blackstart power restoration been more vital for national resilience[1]. At the same time, our capabilities to conduct blackstart have never been at greater risk. This Handbook provides transmission system operators and their partners, including public and private sector organizations, with recommendations to strengthen the survivability of blackstart systems and thereby help ensure societal continuity when catastrophic, Black Sky outages occur.

A growing range of malicious threats and natural hazards can create such Black Sky events, defined as the loss of grid-provided power over multiple

1 For this study, we define Resilience as "The ability to withstand or reduce the magnitude and/or duration of disruptive events, (including events with extreme but plausible catastrophic consequences), which include the capability to identify vulnerabilities and threats, and plan for, prepare for, mitigate, absorb, adapt to, and/or timely recover from such an event.

regions of the U.S. and Canada or equivalent wide-area blackouts on other continents. Without careful, multi-sector preparation, such events would result in cascading disruption or failure of all other societal infrastructures and product and service supply chains, throughout – and beyond – the affected regions. To save and sustain lives when these catastrophic outages occur, it will be essential to conduct blackstart restoration - that is, restarting the electric grid from inside the blacked out area.[2]

Doing so will present extraordinary challenges. The same catastrophic events that necessitate blackstart restoration will also disrupt the operation of blackstart-essential systems, including generators, transmission, and interdependent infrastructure, especially natural gas transmission pipelines and emergency communications networks.

Extraordinary opportunities are emerging to strengthen blackstart resilience. The accelerating deployment of solar, wind, and battery energy storage systems (BESS) could enable transmission system operators to develop innovative strategies for blackstart that leverage the decentralized nature of such generation assets.

At the same time, extraordinary opportunities are emerging to strengthen blackstart resilience. The accelerating deployment of solar, wind, and battery energy storage systems (BESS) could enable transmission system operators to develop innovative strategies for blackstart that leverage the decentralized nature of such generation assets. These assets constitute inverter-based

2 This study uses the term "blackstart power restoration" to encompass the full range of capabilities and functions necessary to restart the grid from inside a blacked-out region without relying on externally provided power, from the initial startup of a blackstart resource through the "next start" process, and re-establishing the reliable operation of bulk electric system (BES) interconnections. North American Electric Reliability Corporation (NERC) defines a blackstart resource as "A generating unit(s) and its associated set of equipment which has the ability to be started without support from the System or is designed to remain energized without connection to the remainder of the System, with the ability to energize a bus, meeting the Transmission Operator's restoration plan needs for Real and Reactive Power capability, frequency and voltage control, and that has been included in the Transmission Operator's restoration plan." NERC, Glossary of Terms Used in NERC Reliability Standards, March 8, 2023, 6, https://www.nerc.com/pa/Stand/Glossary%20of%20Terms/Glossary_of_Terms.pdf. Section B of this Chapter provides an overview of the phases of blackstart restoration.

resources (IBRs). The inverters that help control their power flows could, over time, also enable new ways to restart the grid. That is especially true of BESS that are connected to high voltage transmission systems. Opportunities may also emerge to support power restoration with the millions of IBRs that are connected to electric distribution systems.

At present, however, IBRs can constitute a threat to the grid's reliability when disturbances occur, and are far from ready to lead blackstart restoration. Natural gas-fueled generators will be vital for such restoration for many years to come. Yet, in some regions of the U.S. and other nations, federal and state decarbonization policies are putting increasing pressure on the availability of gas generation. This Handbook proposes options to ensure such availability for blackstart as decarbonization goes forward, and as advances in IBRs and other zero-carbon generation assets (including small modular reactors and geothermal power) enable new restoration plans and capabilities.

However, to restart the grid in the aftermath of a Black Sky event, these assets and other blackstart infrastructure must be able to survive the damage that the event inflicts. The Handbook offers recommendations to strengthen the resilience of such infrastructure, including against electromagnetic pulse (EMP) attacks and the cyber disruption of natural gas infrastructure and other restoration-critical assets and functions.

Key findings and options for consensus-building on blackstart initiatives

A. Finding: **Societal Continuity is Increasingly Dependent On Resilient Blackstart Capabilities**

The foundational importance of electricity to public health and safety, the economy, and national security continues to grow. The accelerating electrification of the transportation systems and other infrastructure sectors, and the tight and complex interdependencies between them, make the sustained availability of power all the more vital. The potential consequences of Black Sky outages will scale accordingly. So, too, will the imperatives to strengthen blackstart resilience against intensifying challenges.

Climate change poses one such challenge. If multi-region power outages occur during increasingly common and severe weather events, including deep, extended freezes and heat domes, blackstart will need to go forward in the face of intensive, competing demands for power restoration to meet human needs, and (especially in the case of natural gas pipelines, electric powered compressor stations and other gas infrastructure) with the disruption of restoration-essential systems in interdependent sectors.

The foundational importance of electricity to public health and safety, the economy, and national security continues to grow.

China and other potential adversaries recognize the importance of blackstart for societal continuity. That makes blackstart infrastructure and operations potential priorities for attack. The implication: blackstart systems must not only be capable of surviving initial EMP, cyber, or combined cyber-physical attacks on the grid that create a wide area outage, but also follow-on attacks that adversaries specially design to disrupt and delay the restoration process.

RECOMMENDATIONS

- **Build on ongoing efforts to map the interdependencies between electricity-dependent infrastructure and adjust power restoration plans accordingly.**

  Transmission system operators (TOPs) already have detailed, carefully vetted plans to prioritize the restoration of service to 1) critical loads that are essential for re-energizing the grid, including electric, natural gas, and communications facilities that support that process; and 2) priority loads that are essential for public health and safety and national security, including water and wastewater systems, hospitals, military bases, and other customers identified in collaboration with emergency managers and other officials.[3]

  TOPs should build on these existing plans to account for new types of priority loads (data centers, etc.) and the impact of electrification on risks of cascading disruptions and the emergence of new single points of failure between multiple sectors. As part of this process, emerging blackstart best practices such as the "Minimum Grid" approach should be carefully examined as options for addressing timeline-critical restoration of essential societal assets.[4]

  To support this planning process, the Department of Homeland Security (DHS) should ramp up its sharing of new data on National Critical Functions (NCFs) that constitute "the functions of government and the private sector that are so vital to the United States that their disruption, corruption, or dysfunction would have a debilitating effect on security, national economic security, national public health or safety."[5]

3 This typology of critical and priority loads draws on PJM, System Restoration, 2020, slides 120-126, https://www.pjm.com/-/media/training/nerc-certifications/trans-exam-materials-2020/transmission-itp/lesson-10-system-restoration.ashx

4 See below, Chapter VII, MEETING RESILIENCE REQUIREMENTS AGAINST BLACK SKY HAZARDS; Section A. PLANNING: 1. Minimum Grid Concept

5 CISA, National Critical Functions, https://www.cisa.gov/topics/risk-management/national-critical-functions#:~:text=National%20Critical%20Functions%20(NCFs)%20are,safety%2C%20or%20any%20combination%20thereof.

In particular, DHS should apply its ongoing NCF Risk Architecture initiative to help blackstart plans account for emerging key assets, networks and systems that support those functions. Artificial Intelligence (AI) tools can help integrate and assess the immense data flows that such efforts will require and provide decision support to TOPs and their government partners to refine restoration plans accordingly. Consistent with the October 2023 Executive Order on the Safe, Secure, and Trustworthy Development and Use of Artificial Intelligence, DOE should collaborate with TOPs to develop and employ AI tools to support power restoration and help manage the risks that AI entails.[6]

- **Develop more robust consultative mechanisms and organizational arrangements to help TOPs shape and refine their restoration plans.**

  Many TOPs, reliability coordinators (RCs), and other participants in the restoration process regularly discuss blackstart planning with emergency managers and other incident response officials. But other transmission system operators and their RCs have less direct connectivity with emergency managers. Given the diversity and complex supporting sub-functions of NCFs that will require restoration, together with traditional restoration priorities that will remain crucial for blackstart planning, it would be useful to broaden and formalize these consultative mechanisms.

  The Handbook proposes specific options for transmission system operators and their partners (including regulators, government officials, interdependent infrastructure operators) to establish coordinating councils that reflect their own service areas and specific needs. A prime option: rely on RTOs/ISOs to help create and coordinate the work of the councils. In regions that do not have an RTO or ISO, that role could be played by other coordination groups (e.g., VACAR and the Georgia

6 The White House, Executive Order on the Safe, Secure, and Trustworthy Development and Use of Artificial Intelligence, October 30, 2023, https://www.whitehouse.gov/briefing-room/presidential-actions/2023/10/30/executive-order-on-the-safe-secure-and-trustworthy-development-and-use-of-artificial-intelligence/

ITS). In addition, the North American Transmission Forum's Resilience Practice Group encourages members to build relationships with the appropriate parties well in advance of emergencies so that coordination does not happen for the first time amid an event. That work provides a valuable basis for further progress.

However, the input that these councils provide should remain advisory. The North American Electric Reliability Corporation (NERC) assigns specific responsibilities to TOPs, RCs, and other subsector entities for blackstart planning, training, and execution, including the responsibilities and functions specified in EOP-005-3 – System Restoration from Blackstart Resources.[7] Accordingly, while coordinating councils can provide crucial input on restoration priorities, TOPs and their NERC-designated partners must continue to have the final say over blackstart planning and execution.

B. Finding: **Blackstart Resilience Planning Must Include All Phases of the Restoration Process**

Many studies of blackstart focus on the initial phases of blackstart: that is, after first assessing the status of blackstart infrastructure and supporting systems (including communications) TOPs will employ "blackstart resources" to begin the restoration process. NERC defines a blackstart resource as:

"A generating unit(s) and its associated set of equipment which has the ability to be started without support from the System or is designed to remain energized without connection to the remainder of the System, with the ability to energize a bus, meeting the Transmission Operator's restoration plan needs for Real and Reactive Power capability, frequency and voltage control, and that

7 North American Electric Reliability Corporation (NERC), EOP-005-3 – System Restoration from Blackstart Resources, effective April 1, 2019, R6, https://www.nerc.com/pa/Stand/Reliability%20Standards/EOP-005-3.pdf.

has been included in the Transmission Operator's restoration plan."[8]

However, starting up blackstart units constitutes only the first phase of a complex, multi-stage restoration process. Once running, blackstart units must then energize transmission facilities to sequentially restart generators along a cranking path, and thereby establish an initial power island within the blacked-out region.[9] TOPs and their partners then add additional "next start" generation (balanced with the pickup of additional load) to grow those islands, maintain their stability, and integrate and synchronize them with neighboring islands until the grid is fully restored.

Starting up blackstart units constitutes only the first phase of a complex, multi-stage restoration process.

Once running, blackstart units must then energize transmission facilities to sequentially restart generators along a cranking path, and thereby establish an initial power island within the blacked-out region.

Each of these phases will confront TOPs and their partners with distinct problems of ensuring the adequacy of generation assets to carry the restoration process forward especially in the face of zero-carbon targets and regulatory initiatives.[10] Grid operators will also face restoration-phase-specific challenges in maintaining operating reliability of their systems. NERC's definition of operating reliability is "the ability of the electric system to withstand sudden disturbances, such as electric short circuits or the unanticipated loss of system elements from credible contingencies, while avoiding uncontrolled cascading blackouts or damage to equipment."[11] In the context of blackstart, however, maintaining the reliability of power islands as they add generation and load will entail special difficulties – especially if adversaries target those operations for disruption.

8 NERC, Glossary of Terms, March 8, 2023, 6, https://www.nerc.com/pa/Stand/Glossary%20of%20Terms/Glossary_of_Terms.pdf

9 NERC, Glossary of Terms, 11

10 Resource adequacy is ability of the electricity system to supply the aggregate electrical demand and energy requirements of the end-use customers at all times, taking into account scheduled and reasonably expected unscheduled outages of system elements. NERC, Reliability Terminology, April 2013, https://www.nerc.com/AboutNERC/Documents/Terms%20AUG13.pdf

11 Ibid

In addition, while (consistent with NERC terminology) the Handbook employs the term "resource adequacy" with regard to assessing potential generation shortfalls, many Handbook readers my find the term "capability" may be more appropriate, given the importance of challenges involving 1) the electrical location of the resource with respect to the next-start generation; 2) the amount of switching that needs to take place to clear the cranking path and connect the resource to it; and 3) the adequacy of fuel supplies for blackstart, rather than focusing on just the output capability of generation assets.

RECOMMENDATIONS

- **Address the resilience challenges (and solutions) for blackstart restoration across the entire process.**

  Ensuring the availability and effective functioning of blackstart resources is necessary but far from sufficient. Unless TOPs, RCs and their partners have adequate next start generation assets and can grow and synchronize their power islands "under fire," starting up blackstart resources and energizing the next bus on the cranking path will be useless. The findings and recommendations that follow address restoration-phase-specific issues for blackstart generation adequacy, operating reliability, and systemic resilience across the full restoration process.

- **Be prepared for adversaries to target emergency communications as a prime attack vector across all restoration stages.**

  TOPs are making concerted efforts to strengthen the survivability of the emergency communications systems they would employ for their own cranking path operations. System-wide restoration improvements must go forward as well. NERC's November 2022 GridEx VI highlighted the imperative to "strengthen operational coordination between the electricity industry and communications providers," and between the broader array of grid operators necessary for growing and synchronizing power islands. The GridEx VI Tabletop scenario featured a widespread loss of landline and cellular communications while electricity utilities were recovering from the cyber and physical attacks and restoring the grid. Exercise participants "agreed that the loss of communications would essentially halt the grid restoration process."[12]

In addition to strengthening operational coordination with communications providers, BES entities responsible for blackstart should also leverage the advances being made by the Rapid Attack Detection, Isolation and Characterization Systems (RADICS) exercise program conducted by the Defense Advanced Research Projects Agency (DARPA) and DOE. Together, they are developing best practices for such exercises and training, as well as the refinement of techniques to clear networks of malware so that restoration can go forward. DARPA, DOE, and industry should partner to widely apply these advances in resilience.[13] DOE should expand their outreach to the electric industry to raise awareness of these exercises and the resilience best practices they are generating. Chapter VIII also proposes specific hardening measures to strengthen emergency communications survivability, including the deployment of the BSX system and other resilient networks. Amateur Radio Emergency Communications groups, using HAM radios and other basic communications tools, should not be overlooked as a component of a resilient, "defense in depth" communications architecture.

12 NERC, GridEx VI Lessons Learned Report, April 2022, vi

13 Defense Advanced Research Projects Agency (DARPA), Technologies to Rapidly Restore the Electric Grid after a Cyberattack Come Online, February 23, 2021, https://www.darpa.mil/news-events/2021-02-23; DARPA, Rapid Attack Detection, Isolation and Characterization Systems (RADICS), https://www.darpa.mil/program/rapid-attack-detection-isolation-and-characterization-systems. This section and other material in the Handbook also draw on RADICS-related findings and recommendations found in Paul Stockton, Blackstart for National Security: Plans, Capabilities, and Exercises for the Emerging Threat Environment, May 19, 2021, a research project funded by the Defense Advanced Research Projects Agency. The views and conclusions contained in this document are those of the author and should not be interpreted as representing the official policies, either expressed or implied, of the defense advanced research projects agency or the U.S. government.

- **Draw on lessons learned from past severe events.**

  While Winter Storm Elliott (December 2022) fell far short of constituting a Black Sky event, it nevertheless highlighted significant problems for ensuring the reliability and availability of black start resources, next start generation, and the natural gas fuel supplies on which they depend.[14] The Handbook examines specific lessons learned from Elliot, Winter Storm Uri (February 2021), and other events that can be scaled up to assess resilience requirements for Black Sky restoration.[15] As severe weather events become increasingly frequent and disruptive, TOPs and their partners should continue to analyze them for blackstart implications.

14 FERC, NERC and Regional Entity Staff Report, Inquiry into Bulk-Power System Operations During December 2022 Winter Storm Elliott, at pp. 105-106 (Oct. 2023), https://www.ferc.gov/media/winter-storm-elliott-report-inquiry-bulk-power-system-operations-during-december-2022

15 FERC-NERC-Regional Entity, Joint Blackstart and Next-Start Resource Availability Study in the Texas Interconnection, December 19, 2023, https://www.ferc.gov/media/ferc-nerc-regional-entity-joint-blackstart-and-next-start-resource-availability-study-texas

C. Finding: **Grid Operators Should Be Ready to Conduct Blackstart in the Midst of Societal Chaos and Intense Political Pressures**

NERC's mandatory standards, regulatory guidelines, and ongoing research initiatives provide a vital framework for strengthening blackstart preparedness. Yet, far beyond the scope of those standards, additional factors could disrupt the restoration process. The most important of these is societal panic. The U.S. Director of National Intelligence warns that "If Beijing feared that a major conflict with the United States were imminent, it almost certainly would consider undertaking aggressive cyber operations against U.S. homeland critical infrastructure," aimed at "inducing societal panic" and achieving related strategic goals.[16] Other adversaries could also attack the grid to incite public disorder in the U.S. and its global security partners, especially if those attacks were paired with information operations to further inflame panic.

These societal effects are unlikely to impede the earliest phases of blackstart. However, as restoration personnel deploy to increasingly far-flung substations and other critical locations, and the disruptive effects of the blackout create intensifying threats to water and other basic human needs, the loss of public order could put those personnel at increasing risk. An extended outage would threaten continuity of government and governmental functions at all levels which will also create intense political pressure on elected officials at all levels of government to "help" the restoration process and insist on prioritizing restoration to specific customers in ways that ignore engineering constraints or broader needs, such as prevention of cross-sector critical infrastructure failures.

16 Office of the Director of National Intelligence, Annual Threat Assessment of the U.S. Intelligence Community, February 6, 2023, 10, https://www.dni.gov/files/ODNI/documents/assessments/ATA-2023-Unclassified-Report.pdf

RECOMMENDATIONS

- **Accelerate exercises and planning for Grid Security Emergencies.**

The Federal Power Act (FPA) grants the Secretary of Energy far-reaching emergency authorities that could help guide and prioritize blackstart restoration in severely disrupted environments. If the president declares that a cyberattack or other attack or disaster has created a grid security emergency (GSE), the Secretary of Energy has the authority to issue GSE orders to "protect or restore" the reliability of defense critical electric infrastructure (DCEI) – that is, any infrastructure in the continental United States that services critical defense facilities.[17] Multi-region blackouts created by China, Russia, or other potential adversaries would almost certainly constitute a GSE. Prioritizing the restoration of service to critical defense facilities would be of paramount importance in such a conflict. However, restoring service to other critical/priority loads would remain crucial as well, including water systems and other electricity-dependent infrastructure essential for protecting society and public order.

> Government leaders and industry should develop criteria in advance of Black Sky events to help allocate and sequence the delivery of scarce security resources, rather than make up such guidelines in the midst of the catastrophe.

DOE, TOPs, and other blackstart partners should prepare in advance for GSEs by 1) developing "template" GSE orders that account for national security and other critical/priority loads in a given TOP service area, which DOE would modify as needed to fit the circumstances of an actual emergency; 2) exercise the issuance and implementation of such orders; 3) ensure that survivable communications are available to enable

17 The FPA states that the Secretary shall designate CDFs in consultation with other appropriate Federal agencies and with the owners and operators of infrastructure that may be DCEI. PL 114-94, p. 1773

the delivery of orders and subsequent messaging; and 4) address other issues of concern to these partners, including indemnification of utilities against the unintended consequences of order compliance.

- **Scale up planning for law enforcement support.**

  Police and State National Guard (NG) personnel routinely provide public safety and security support to utilities during restoration operations that follow severe hurricanes and other natural disasters. However, Black Sky outages will create competing demands on these resources to secure a wide range of facilities and maintain public order. Government leaders and industry should develop criteria in advance of Black Sky events to help allocate and sequence the delivery of scarce security resources, rather than make up such guidelines in the midst of the catastrophe. Building on existing plans to deal with less severe events offers an opportunity to facilitate such collaboration.

  Support will also be needed from government agencies apart from law enforcement. For example, in the early days of COVID-19, concerns emerged regarding the ability of essential TOP personnel to travel to work, or to travel as needed in performance of their duties, amid stay-at-home orders. In some cases, TOP entities sought authorization from state governors to ensure freedom of movement. In a Black Sky event, freedom of movement could be even more restricted, and TOPs may need government assistance to obtain resources and move material necessary for power restoration. TOPs should engage with DOE and state, local, tribal, and territorial (SLTT) agencies to identify potential support requirements and develop plans to meet them.

- **Build on the Existing Playbook System to Help Counter Enemy Information Operations.**

  The best way to limit the psychological impact of wide-area blackouts is to rapidly restore power. However, given the disruptive effects that information operations (IOs) could have during the restoration process, measures to counter such IOs will also be crucial. The Electricity Subsector Coordinating Council (ESCC) and its members have built an effective, frequently used system to coordinate public communications when blackouts or other major grid disturbances occur. Electric utilities and their industry and government partners should build on these playbooks to explicitly account for chaotic conditions in which blackstart restoration will likely need to go forward, and (together with government leaders) coordinate counter-messaging against adversary disinformation.

D. Finding: **Dangers to Blackstart Resource Adequacy Are Growing**

The accelerating retirement of coal and (in some regions) natural gas-fueled generators confront TOPs and their partners with far-reaching challenges for maintaining the grid's reliability. In particular, as fossil retirements accelerate, risks are intensifying to both the adequacy of generation resources and the grid's ability to withstand disturbances and limit their systemic effects – including from Black Sky events. Yet, advances in inverter technologies are now creating opportunities to mitigate these problems and develop new strategies to strengthen grid resilience. Probabilistic assessment of asset vulnerabilities and consequences could also help identify and mitigate critical points of potentially catastrophic failures.[18]

18 National Academy of Engineering, Creating A Sustainable National Electric Infrastructure While Maintaining Reliability and Resiliency of the Grid. February 2023. Report available at https://www.nae.edu/289966/Public-Documents

The transformation of the U.S. generation fleet poses an equivalent mix of risks and opportunities for blackstart restoration. Already, Regional Transmission Operators (RTOs) such as PJM are proposing to allow hybrid resources (e.g., solar and wind generation paired with energy storage) to serve as "fuel assured blackstart units."[19] FERC did accept PJM's specific proposal for these hybrid resources as "just and reasonable". Yet, as will be detailed later in this Executive Summary, it will take many years – and the resolution of immense technical and operational problems - before IBRs can replace gas-fired generators for next start and subsequent restoration phases.

Federal and state zero-carbon policies and regulatory measures will jeopardize the availability of gas generators. NERC has found that environmental rules, together with the costs of keeping blackstart resources compliant with NERC's critical infrastructure protection standards and other factors, are already driving down the availability of these resources.[20] Decarbonization will also put increasing risk on gas next-start generators, and on the investments in pipeline systems and other gas infrastructure necessary to ensure the resilient flow of fuel to these assets. Yet, despite these dangers and the strategic imperative for resilient restoration capabilities, no current Federal or state decarbonization initiatives directly address the threats that zero carbon targets pose to blackstart.

---

19 PJM, PJM Interconnection, L.L.C., Docket No. ER23-1874-000 Black Start Service Revisions to Enhance Fuel Assurance, Filing to the Federal Energy Regulatory Commission, May 12, 2023, 20, https://www.pjm.com/directory/etariff/FercDockets/7345/20230512-er23-1874-000.pdf. Note that the PJM's proposal is structured to provide opportunities for the use of what it terms "intermittent" resources ("wind, solar, and run-of-river hydroelectric resources") and "hybrid" resources ("intermittent resources paired with storage"). PJM, Affidavit of Michael E Bryson on Behalf of PJM Interconnection LLC, May 12, 2023, 47, https://www.pjm.com/directory/etariff/FercDockets/7345/20230512-er23-1874-000.pdf. Section D of this study provides a detailed analysis of PJM's tariff proposal and its implications for future blackstart system planning and capabilities in greater detail.

20 NERC, CIP-002-6, Cyber Security: BES Cyber System Categorization, Supplemental Material, 35, https://www.nerc.com/pa/Stand/Reliability%20Standards/CIP-002-6.pdf

RECOMMENDATIONS

- **Leverage emerging best practices in state legislation and regulations that support grid reliability and resilience.**

Twenty-two states have enacted goals to achieve 100% carbon-free electric generation, with more considering adoption of such targets. However, some of these states, including New York and Virginia, have recently adopted special provisions to enable their decarbonization programs to change if necessary to protect the reliability and security of electric service.[21] Regulators, legislators, and their industry partners should develop equivalent, specifically-targeted measures to ensure the adequacy of blackstart generation assets, including options for "carveouts" that protect such assets (and potentially critical load units as well) as deep decarbonization goes forward.

Regulators, legislators, and their industry partners should develop equivalent, specifically-targeted measures to ensure the adequacy of blackstart generation assets, including options for "carveouts" that protect such assets (and potentially critical load units as well) as deep decarbonization goes forward.

However, establishing carveouts for black start resources under state policies will entail significant problems. Blackstart units need to be electrically close to next start generation for the black start unit to crank. Ensuring the availability of such adjacent assets could require broadening the scope (and both cost and carbon footprint) of carveout initiatives. Moreover, even if both the black start and next start generation remain 'available' if these units are not run routinely, staffed, maintained, etc. they will not be reliable in a restoration event.

21 See, for example, Code of Virginia, § 56-585.5. Generation of electricity from renewable and zero carbon sources. (2023 updated section) 2, https://law.lis.virginia.gov/vacodeupdates/title56/section56-585.5/

The Handbook also proposes options to address the problems posed by CIP compliance costs and other disincentives for generator owners to continue to provide blackstart services. One promising option: adopt a phased approach, in which resources identified as essential would be eligible for specialized cost recovery that would include maintenance, and then transition those assets for other uses (even, potentially, to operate carbon capture systems).[22] Ultimately, however, if state policies require total decarbonization, achieving that goal for blackstart will require the deployment of grid forming IBRs and associated control systems at scale – an enormous technical and operational challenge.

CIP compliance costs are not the only source of downward pressure on resource availability. EOP standards, including EOP-005-3, System Restoration from Blackstart Resources, also entail such costs. These standards are valuable for blackstart resilience: they increase the likelihood resources will function when called upon. However, cost recovery for compliance with EOP standards must also be part of broader measures to ensure resource adequacy.

22 DOE, Carbon Capture Opportunities for Natural Gas Fired Power Systems, https://www.energy.gov/fecm/articles/carbon-capture-opportunities-natural-gas-fired-power-systems; International Energy Agency, CO2 Capture at Gas Fired Power Plants, July 2012, https://www.globalccsinstitute.com/archive/hub/publications/103211/co2-capture-gas-fired-power-plants.pdf

E. Finding: **Immediate Initiatives Are Required to Ensure the Operating Reliability of Blackstart in an IBR-Heavy Grid**

As with resource adequacy, the ongoing revolution in the generation mix is creating challenges for the operating reliability of the BES as a whole. And again, as with adequacy, the transformation of the generation fleet will create especially severe challenges to the operating reliability of blackstart systems across the United States and in partner nations.

The BES is now what can be termed as a "hybrid electric system" – that is, it comprises both inverter-based resources and large synchronous generators that have traditionally supported grid reliability, including fossil, hydro, and nuclear power plants.[23] In theory, IBRs can make significant contributions to blackstart restoration. Leveraging the emerging capabilities of battery energy storage systems (BESS) and other generation sources (See Chapter VII) for restoration could be especially valuable. In practice, however, IBRs currently pose a persistent threat to grid reliability, and – potentially – to the ability of grid operators to maintain the stability of the blackstart power islands they seek to establish and grow in Black Sky outages.

RECOMMENDATIONS

- **Take stronger measures to remedy the performance failures of IBR.**

  In theory, the voluntary standards issued by Institute of Electrical and Electronics Engineers (IEEE) will remedy the tendency of inverters to trip off during disturbance, halting the flow of power from the solar and wind assets they control. Such standards now exist for IBRs connected to the BES as well as to distribution-level systems (i.e., distributed

23 For a characterization of this hybrid system, see Yashen Lin et al, Research Roadmap on Grid-Forming Inverters, National Renewable Energy Laboratory (NREL), November 2020, iv, https://www.nrel.gov/docs/fy21osti/73476.pdf

energy resources, or DERs). In practice, severe IBR performance failures continue to occur. NERC is conducting Project 2023-02 on Performance of IBRs to ensure generation owners are analyzing and mitigating abnormal performance issues when they occur. NERC is also calling for the development of a comprehensive ride-through standard to replace NERC PRC-024-3.[24] As these important efforts go forward, NERC should also address the associated IBR trip-off problems (and benefits for improved ride-through) for restoration, and the broader problems this Handbook identifies for conducting blackstart in an IBR-heavy grid.

Immediately address the danger of adversary mis-operation of IBRs. DOE warns that "malicious actors are well-positioned to enter DER energy systems.[25] Many of the same vulnerabilities will apply to IBRs connected to bulk power system. In particular, with the deployment of IEEE-compliant inverters that can help maintain frequency and voltage control, the ability of adversaries to access and mis-operate these devices could have catastrophic consequences. Those dangers apply both to adversary efforts to create Black Sky outages and to the disruption of the restoration process in areas with very high IBR penetration. NERC, FERC, DOE, the National Association of Regulatory Utility Commissioners (NARUC) and their industry partners should continue to advance options to mitigate these risks, including initiatives to ensure "security by design" for such devices, and explore options to establish mandatory, enforceable requirements for the deployment of inverters that meet these standards.

24 NERC, Southwest Solar, 6

25 DOE, Cybersecurity Considerations for Distributed Energy Resources on the U.S. Electric Grid, October 2022, 16, https://www.energy.gov/sites/default/files/2022-10/Cybersecurity%20Considerations%20for%20Distributed%20Energy%20Resources%20on%20the%20U.S.%20Electric%20Grid.pdf

F. Over the Longer Term, Ibrs Could Offer Unique Benefits to Blackstart Resilience – But Only If Their Technical and Operational Problems Can Be Resolved

Significant challenges remain to replacing gas-fired black start units with hybrid plants. For example, if the batteries in those plants are to serve as blackstart units, their stored power must be reserved for that function. Batteries will be useless for energizing cranking paths if their power has already been discharged for normal electricity dispatch and market operations. Specialized financial incentives and associated contract terms of service will be essential to help asset owners recoup their investments in hybrid plant blackstart units and incentivize the deployment of BESS with long-duration storage capabilities. Cost recovery measures may also be needed to ensure that selected batteries reserve a portion of their power so that is always available for blackstart restoration,

If BESS or other generation sources can provide an alternative to gas generation-dependent restoration systems, including as blackstart resources, cranking generators, and next start assets, potential adversaries will be less able to disrupt restoration by cutting off the flow of natural gas.

However, assuming that hybrid plants can meet these and other functional requirements, IBRs could offer significant benefits for Black Sky resilience as well. The most obvious is diversifying the types of generation. If BESS or other generation sources can provide an alternative to gas generation-dependent restoration systems, including as blackstart resources, cranking generators, and next start assets, potential adversaries will be less able to disrupt restoration by cutting off the flow of natural gas.

Employing IBRs as black start units could also increase the redundancy of blackstart cranking paths. While each black start unit can only crank one Critical Load at a time, it is possible for multiple units to energize separate cranking paths and power islands in parallel.[26] The ability to do so would help mitigate the risk that the initial Black Sky event that requires blackstart restoration will damage the generation or transmission infrastructure of one or more cranking paths (including, potentially, the primary path). Redundant paths can also strengthen system resilience if adversaries specifically target restoration infrastructure with either cyber or combined cyber-physical attacks.

RECOMMENDATIONS

- **Build on lessons learned from PJM's initiatives on blackstart and IBRs.**

  FERC has approved the tariff proposal by PJM on Black Start Service Revisions to Enhance Fuel Assurance, which sets criteria for selecting IBRs as and other types of generation (including gas-fired generators) as "Fuel Assured Black Start Units."[27] As the implementation of that effort goes forward, other BPS entities will be able to draw valuable lessons on equivalent measures modified to meet their own requirements and circumstances.

- **Develop multi-island restoration strategies that leverage accelerating hybrid plant deployments.**

  Depending on the topography of transmission and distribution infrastructure and the connectivity they provide to critical and priority loads, the use of widely distributed IBR units could help TOPs establish multiple power islands simultaneously in ways that enable much faster restoration timelines than would be possible with single-island plans. Such strategies could also strengthen system redundancy against the targeted destruction of cranking path transmission lines, substations, and other infrastructure. Efforts to develop multi-islanding restoration strategies should be accompanied by training to teach operators to

26 PJM, Black Start Service Revisions, 20-22

27 PJM, Black Start Service Revisions, 17-118

operate a grid of small islands (which is a specialized skill) and bring the islands together.

- **Rethink the tradeoffs that CIP standards strike between the need to ensure resource adequacy, versus the imperative for generation survivability.**

  Local, state, and federal initiatives to reduce carbon emissions (and in some cases mandate a zero-carbon grid) provide only one of the downward pressures on the availability of gas-fired generators for blackstart restoration. Ensuring adequate compensation and cost recovery for generation owners in organized markets will also be crucial. The costs of complying with cyber and physical security regulations, and the risk that generation owners will need to pay penalties for non-compliance, creates further availability issues. Yet, given the intensifying threats to the grid and danger that adversaries prioritize the disruption of blackstart resources and operations, we also need to examine whether regulations for such resources are sufficiently robust and comprehensive. Thermal generation retirements impact critical load units, which are cranked by the blackstart units. If grid operators and their partners seek to expand their definitions of critical/priority loads, they will need sufficient critical/priority generation as well.

  NERC's Supplemental Material for CIP-002-6 - Cyber Security - BES Cyber System Categorization highlights the competing pressures that regulators and their partners face in ensuring both the adequacy and survivability of blackstart resources. Going forward, NERC and its partners should reassess this balance in light of the intensifying threats to restoration assets from China and other potential adversaries. They should also develop new funding strategies to compensate generator owners for higher levels of security than required under a low impact rating and increase the financial incentives for GOs to continue providing blackstart services even if their compliance costs grow.

G. Finding: **Grid Following and Grid Forming Inverters Pose Distinct Problems – and Different Potential Advantages – for Blackstart Resilience**

Above and beyond the ride-through and other emerging capabilities of IBRs, advances in the ability of inverters to "form" the grid in the aftermath of power outages may also help these resources contribute to blackstart, and – potentially – enable the development of new restoration strategies that provide greater resilience against Black Sky threats.

DOE summarizes these opportunities. To restart the grid after a blackout in the current hybrid system, grid operators must first start up synchronous generators before they can connect IBR-generated power to the grid. The reason: when a large disturbance or outage occurs on the grid, conventional inverters will shut off power to these energy sources and wait for a signal from the rest of the grid that the disturbance has settled and it is safe to restart—known as "grid-following" (or GFL). As wind and solar account for increasing shares of the overall electricity supply, it is becoming impractical to depend on the rest of the grid to manage disturbances. Grid-forming inverters (or GFM) are an emerging technology that allows solar and other IBR energy sources to restart the grid independently.[28]

As wind and solar account for increasing shares of the overall electricity supply, it is becoming impractical to depend on the rest of the grid to manage disturbances. Grid-forming inverters (or GFM) are an emerging technology that allows solar and other IBR energy sources to restart the grid independently.

Both types of inverters may potentially enable TOPs to supplement or ultimately replace synchronous generators for restoration, and further exploit

28 DOE, Powering On with Grid-Forming Inverters, January 4, 2021, https://www.energy.gov/eere/solar/articles/powering-grid-forming-inverters

the resilience advantages of the wide dispersal of generation assets. However, as the Handbook discusses in detail, the deployment of GFL resources at scale will pose multiple problems for keeping power islands stable, load pickup, maintaining the effectiveness of protection systems, and other crucial issues. Resolving the problems of using GFMs to supplant synchronous generation will also be difficult. Understanding of GFM controls and their impact on BES performance is still in the early stages, and the operational difficulties of controlling their functions will present novel problems. Nevertheless, as NERC concludes, GFM technology "shows significant promise."[29]

RECOMMENDATIONS

- **Act now to ensure battery energy storage systems (BESS) and hybrid plants (which combine IBR generation and storage) have built-in capabilities to support restoration.**

  While GFM controls can potentially be implemented on any type of IBR including new solar and wind plants, batteries are the low-hanging fruit. GFM behavior requires a certain amount of energy buffer, which for wind and solar resources means they must continuously operate below their maximum available power production. In addition, GFM control in wind turbines may result in greater and more frequent mechanical stress. But utility scale batteries have none of those drawbacks. The battery is the energy buffer, and only software modifications to a battery's controls are needed to make the battery a GFM resource. But ensuring those software features are actually adopted poses a significant challenge. Regulators, hybrid plant operators, and other stakeholders should develop incentives (including non-market options) for the deployment of restoration-capable BESS, rather than try to bolt the necessary software and control systems after the battery fleet is built out.

29 NERC, White Paper: Grid Forming Technology: Bulk Power System Reliability Considerations, December 2021, https://www.nerc.com/comm/RSTC_Reliability_Guidelines/White_Paper_Grid_Forming_Technology.pdf

- **Incorporate blackstart capabilities in the development of advanced IBR performance standards.**

  A DOE-supported initiative is underway to develop universal performance standards for grid-forming inverters. The Universal Interoperability for Grid-Forming Inverters (UNIFI) Consortium has developed a preliminary version of specifications for these devices, including, potentially, those that would enable IBR support for blackstart and system restoration. However, the Consortium notes that not all GFM IBRs are expected to be able to provide black-start services, since those services may require additional hardware, design, and functionalities. The Consortium should reach out to regulators and other stakeholders to develop specialized cost recovery mechanisms for TOPs and other grid operators that wish to deploy these more expensive (but more capable) GFMs.

H. Finding: **Measures to Secure the Flow of Fuel for Gas-Fired Generators Must Scale With Intensifying Threats**

It will be many years, if ever, before inverter-based resources can replicate the restoration capabilities provided by gas generators. Accordingly, ensuring the survivability of these essential gas generation assets and the infrastructure that delivers fuel to them must remain for societal continuity and national security.

It will be many years, if ever, before inverter-based resources can replicate the restoration capabilities provided by gas generators. Accordingly, ensuring the survivability of these essential gas generation assets and the infrastructure that delivers fuel to them must remain for societal continuity and national security.

Gas transmission system owners and operators continue to take steps to strengthen the resilience of their infrastructure and operations. Going forward, additional measures and system-wide adoption of them will be required to meet emerging threats. The U.S. intelligence community emphasized that Chinese attack capabilities are especially dire. To ensure blackstart resilience, we need not only more robust hardening efforts to stay ahead of the People's Liberation Army, but also much closer collaboration between gas and electric system operators to sustain blackstart when pipeline disruptions occur.

RECOMMENDATIONS

- **Build gas-electric information sharing protocols and communications capabilities to support restoration.**

  New Federal regulations mandate incident reporting to the government. In addition, however, utilities need prompt, assured notification of pipeline disruptions to assess their potential impact on blackstart plans and develop contingency options. Real-time information on linepacks[30] and expected rates of decline will also assist adaptive planning, and will be essential for startup and/or system support from non-gas fueled generation when needed. Utilities, gas companies, and their partners should assess the adequacy of existing "red phone" notification systems and protocols to support blackstart, and ensure the resilience of those systems against adversary attack. Understanding real-time capabilities of blackstart resources during tight gas conditions presents challenges as well. for that reason and many others, it is crucial to include TOPs in information sharing initiatives as well to oversee cross zonal blackstart status. FERC Order No. 787 provides a foundation for progress. Now is the time to build on that Order in ways that also ensure the survivability of gas-electric emergency communications.

30 "Linepack describes the total of volume of gas contained within the system." National Gas, What is Linepack?, www.nationalgas.com/sites/gas/files/documents/Line%20Pack%20Presentation%20-%20Dec%202018.pdf

- **Prioritize resilience measures for secondary fuels.**

  The ability of dual-fuel generators to run on diesel instead of natural gas bolsters blackstart system resilience -- but only if fuel distribution companies can resupply them. In a wide area backout, fuel providers will face overwhelming requests for diesel deliveries in a severely disrupted environment. Moreover, even in normal conditions, many generators will only be able to receive deliveries sufficient for several hours of operations per day. A wide-area blackout will greatly impede delivery operations. Utilities and their ONG and government partners should develop options to meet these challenges, including emergency waivers of transportation and emissions regulations to facilitate secondary fuel operations for blackstart.

  These partners should also intensify their efforts to develop LNG storage options to support restoration in addition, they might and explore whether cost recovery measures and specialized environmental wavers might facilitate increased on-site secondary fuel storage dedicated to support blackstart generation.

- **Begin developing mandatory reliability and infrastructure protection standards for natural gas infrastructure before a Black Sky event makes them inevitable – and too late.**

  In the case of BES reliability and CIP standards, it took a catastrophic outage before the electricity subsector – with help from Congress – transitioned from voluntary to mandatory, enforceable requirements. Gas system operators now have the opportunity to lead the development of such standards, supported by specific Intelligence Community findings on attack vectors and mitigation options.

  An appropriate starting point: establish requirements to prioritize the supply of gas to blackstart generation assets in major outage events. In Black Sky events where available gas supplies are limited, it may be necessary to devote that fuel to blackstart and next start generation assets until restoration is well underway. This could divert supplies from other critical uses; yet, failing to mandate the allocation of fuel to such assets could cripple restoration.

- **Natural gas-fed Blackstart and Next-start units in the U.S. should adopt a 2-layer strategy in protecting their fuel sources.**

  First, a secondary/backup fuel type such as fuel oil must be available on site to use at those generators in the event of natural gas pipeline disruption. Cost recovery measures will need to enable compliance with such requirements to avoid further downward pressure on resource availability. Second, public and private sector investment will be needed to maintain the reliability of gas pipelines to fuel blackstart generation – a significant challenge in states that require decarbonization.

> Thousands of small hydro generators are already in place in the U.S. and could be used to provide additional blackstart capability and overall electric system resilience at all levels: transmission, sub-transmission, and distribution.

I. Finding: **New Hydropower, Geothermal and Fuel Cell Initiatives Can Help Meet Emerging Blackstart Requirements**

In addition to large hydropower units that have traditionally provided blackstart services, smaller units can provide additional blackstart capability if combined with energy storage, and "bottom up" (sub-transmission, distribution) restoration strategies. Thousands of small hydro generators are already in place in the U.S. and could be used to provide additional blackstart capability and overall electric system resilience at all levels: transmission, sub-transmission, and distribution. Additionally, thousands of existing non-powered dams distributed across the U.S. could be repowered and utilized to support distributed electric system resilience.

While the viable geographic locations of geothermal energy production are limited, the plants that do exist are - in principle - all capable of blackstart operations. Advances in drilling technology, coupled with the desire for renewable electricity generation, will allow for development of additional geothermal power plants in the U.S. and throughout the world. Fuel cells are also blackstart capable – the only current limitation on using them for restoration is their size.

RECOMMENDATIONS

- **Establish an interagency, public-private task force to identify promising opportunities for small hydro blackstart generation.**

  FERC has far-reaching statutory authorities over hydropower licensing and overseeing non-Federal hydropower in the U.S.. In addition, DOE is conducting an array of advanced hydropower technology development programs, including some with potential blackstart applications.[31]

31 DOE, Hydropower Technology Development, https://www.energy.gov/eere/water/hydropower-technology-development#:~:text=Hydropower%20Technology%20Development%20%7C%20Department%20of%20Energy

None of these options represent a "silver bullet." Hydro is location and geography dependent. In many regions, for example, pumped storage hydro will never reach the scale necessary to meet blackstart criteria. And while aggregating smaller hydro plants provides an alternative, it will be difficult in most cases to tie them to the cranking paths necessary to serve the critical/priority loads.

- **FERC and DOE should partner with industry to create specially focused initiatives to leverage these opportunities on a nationwide basis. Equivalent efforts should be launched by other nations as well.**

- **Existing and planned geothermal plants of sufficient size and in viable locations should be included as resources in blackstart planning.**

- **As Fuel Cell arrays become larger, they should be considered a blackstart asset, first at microgrid or distribution grid level, and eventually for the BES.**

  The widespread deployment of fuel cells could increase the number of conventional plants that are viable as blackstart units. However, because current-generation fuel cells typically are inverter-based resources, they are subject to some of the same limitations (at least in the short to medium term) as other IBRs .

J. Finding: **Existing Nuclear Power Plants Present Critical Black Sky Challenges, But Could Become Future Black Sky Blackstart Assets**

Given the Black Sky power outage operational challenges for all entities and sectors, Nuclear Power Plants (NPPs) going offline and moving to "safe shutdown mode" could lead to precarious situations. Currently, emergency backup power to run active cooling, control, and safety systems of NPPs is supplied chiefly by diesel generators when they go offline and move into safe shutdown mode.

Black Sky power outages could last from weeks to months and would include cascading failures of other critical infrastructures. As such, the Nuclear Regulatory Commission (NRC) requirement of a 7-day onsite storage of diesel fuel will be inadequate, and the 30-day diesel fuel resupply requirement would most likely be difficult or perhaps even impossible to execute when disruptions of transportation, communication, and even refining of diesel fuel are hindered. For the U.S. NPP fleet, enhanced flexible operations capabilities, as well as plant islanding and blackstart capabilities could enhance plant safety and security, while allowing the NPPs to perform useful power system restoration functions. In practice, major licensing, technical, and operational barriers preclude such use today. In addition to the existing NPPs, the deployment of small modular reactors (SMRs) could open the door to new blackstart applications, but only if such capabilities are engineered into the SMR development and licensing process.

RECOMMENDATIONS

- **The U.S. government should review the licensing, regulations, and any technical barriers currently preventing the existing nuclear fleet from becoming Black Sky assets through more flexible operation and ensure that SMRs are designed to enable plant island mode and blackstart operations.**

- **The nuclear industry trade groups, Nuclear Energy Institute (NEI) and the Institute of Nuclear Power Operations (INPO) should review this document for actions to improve the resiliency of U.S. nuclear power plants. This action could be in the form of a joint task force to develop the Islanding Mode and Blackstart capability at the nuclear generation units to support BES grid restoration.**

1

CHAPTER ONE

BLACKSTART IN A TRANSFORMING GRID

INTRODUCTION: **THE NATURE OF THE CHALLENGE**

Blackstart power restoration - that is, the ability to restart the electric grid from inside a blacked out area[32] - is vital for societal continuity. In Black Sky events that create extensive, multi-region disruptions of grid infrastructure,

32 This study uses the term "blackstart power restoration" to encompass the full range of capabilities and functions necessary to restart the grid from inside a blacked-out region without relying on externally provided power, from the initial startup of a blackstart resource through the re-establishing the reliable operation of bulk power system (BPS) interconnections. North American Electric Reliability Corporation (NERC) defines a blackstart resources as "A generating unit(s) and its associated set of equipment which has the ability to be started without support from the System or is designed to remain energized without connection to the remainder of the System, with the ability to energize a bus, meeting the Transmission Operator's restoration plan needs for Real and Reactive Power capability, frequency and voltage control, and that has been included in the Transmission Operator's restoration plan." NERC, Glossary of Terms Used in NERC Reliability Standards, March 8, 2023, 6, https://www.nerc.com/pa/Stand/Glossary%20of%20Terms/Glossary_of_Terms.pdf. Section B of this Chapter provides an overview of the phases of blackstart restoration.

resilient blackstart capabilities will be essential for restoring power to several critical infrastructure systems, such as water systems, telecommunications, and other community lifelines.[33] Such capabilities will also be vital to ensure that National Critical Functions (NCFs) and military base operations can be sustained.[34] Furthermore, with the accelerating electrification of multiple infrastructure sectors, and the interdependencies and risks of cascading failures between them, the imperative for resilient blackstart restoration capabilities continues to grow.

In Black Sky events that create extensive, multi-region disruptions of grid infrastructure, resilient blackstart capabilities will be essential for restoring power to several critical infrastructure systems, such as water systems, telecommunications, and other community lifelines.

Intensifying natural and manmade hazards both magnify the need for blackstart resilience and put it at risk. Severe weather events highlight this dual challenge. In Winter Storm Uri (February 2021) and Winter Storm Elliott (December 2022), disruptions in the flow of natural gas to gas-fired generators jeopardized the reliability of the bulk power system and – especially in Uri – risked creating wide-area power outages that could have necessitated blackstart restoration.[35] Yet, the same gas supply interruptions also jeopardized

33 The Federal Emergency Management Agency (FEMA) designates community lifelines as the most fundamental services in the community that, when stabilized during or after an event, "enable all other aspects of society to function, " including " the continuous operation of critical government and business functions and is essential to human health and safety or economic security." FEMA, Community Lifelines, https://www.fema.gov/emergency-managers/practitioners/lifelines. Black Sky events are natural or manmade catastrophes that create electric outages across multiple regions of a continent, such as the disruption of one or more interconnections in North America.

34 Department of Homeland Security (DHS) designates National Critical Functions as the "functions of government and the private sector so vital to the United States that their disruption, corruption, or dysfunction would have a debilitating effect on security, national economic security, national public health or safety, or any combination thereof." DHS, Cybersecurity and Infrastructure Security Agency (CISA), National Critical Functions, https://www.cisa.gov/national-critical-functions

35 FERC, NERC and Regional Entity Staff Report, October 2023, Inquiry into Bulk-Power System Operations During December 2022 Winter Storm Elliott, November, 7, 2023, https://www.ferc.gov/media/winter-storm-elliott-report-inquiry-bulk-power-system-operations-during-december-2022; Federal Energy Regulatory Commission (FERC) and North American Electric Reliability Corporation

the ability of blackstart generators to function. The broader implication for blackstart preparedness: the events that disrupt energy assets and cause wide-area outages can also disrupt the operation of blackstart-essential systems, including generators, transmission, and interdependent infrastructure - especially natural gas transmission pipelines and emergency communications networks.

1. Accounting for Natural and Manmade hazards

Building resilience against these twin challenges will be all the more important as climate change continues. In reviewing the impact of Elliott, the Federal Energy Regulatory Commission (FERC) and North American Electric Reliability Corporation (NERC) noted that "the event is the FIFTH in the past 11 years in which unplanned cold weather-related generation outages jeopardized bulk-power system reliability."[36] More frequent and intense droughts will also affect the availability of hydropower to meet hot-weather spikes in demand, and – if necessary – to conduct blackstart under such extreme conditions.[37]

> During Elliot, blackstart-designated generating units totaling 19,000 MW experienced forced outages, derates, or failures to start during the event.

Severe weather can also jeopardize the availability of fossil-fueled blackstart generating units. During Elliot, blackstart-designated generating units totaling

(NERC), Presentation: FERC-NERC-Regional Entity Joint Inquiry Into Winter Storm Elliott, September 21, 2023, https://www.ferc.gov/news-events/news/presentation-ferc-nerc-regional-entity-joint-inquiry-winter-storm-elliott; FERC, NERC and Regional Entity Staff Report, The February 2021 Cold Weather Outages in Texas and the South Central United States, https://www.ferc.gov/media/february-2021-cold-weather-outages-texas-and-south-central-united-states-ferc-nerc-and. See also Cathleen Morehouse, "ERCOT Narrowly Avoided 'Much More Devastating' Impacts as Nearly Half of Generation Went Offline: CEO, Utility Dive, February 25, 2021, https://www.utilitydive.com/news/ercot-narrowly-avoided-much-more-devastating-impacts-as-nearly-half-of-ge/595701/

36 FERC-NERC-Presentation-Inquiry on Winter Storm Elliott, 3, emphasis in the original

37 Minghao Qiu, et al, "Drought impacts on the electricity system, emissions, and air quality in the western United States," Proceedings of the National Academy of Sciences (PNAS), July 6, 2023, https://www.pnas.org/doi/10.1073/pnas.2300395120

19,000 MW experienced forced outages, derates, or failures to start during the event. A November 2023 FERC/NERC report found:

Altogether, 155 blackstart-designated generating units (119 of which were natural gas-fired) reported more than 248 discrete outages, derates or failures to start. Of these, 29 percent reported multiple outages, and 23 percent were start-up failures—i.e. units that failed to perform the essential function of blackstart units. Blackstart generation loss unit types included natural gas-fired, dual-fuel capable, and other primary fuel types.[38]

> Precisely because blackstart power restoration is so crucial for societal continuity and national security, attackers may prioritize the disruption of blackstart infrastructure to help achieve their goals in the conflict and continue attacking the restoration process until they prevail or are defeated.

Given the severity of the threats to blackstart generation illuminated by such severe weather events, FERC and NERC recommended that NERC and its Regional Entities, with Commission staff engagement, should initiate a study "to assess readiness of other blackstart units to operate during cold weather conditions, including recommended actions for improvement, (if necessary)."[39] Subsequent portions of this Handbook recommend options to contribute to that analysis, and also account for a broader range of hazards.[40]

In particular, regulators, TOPs, and their partners should also account for manmade threats to blackstart, which can both necessitate restoration and make it vastly more difficult to perform. The Handbook examines ongoing upgrades in adversary cyberattack capabilities, especially "living off the land" attack vectors, which unless mitigated could catastrophically disrupt grid operations and the flow of fuel to gas-fired generators.[41] EMP strikes can cause Black Sky outages as

38 FERC/NERC, Inquiry into BPS Operations During Elliott, November 2023, 105-6

39 FERC/NERC, Presentation-Inquiry on Winter Storm Elliott, 17

40 To include GMD and HEMP, as detailed in E PRO Handbook IV: EMP: Best Practices and Cost Effective EMP Protection Options for Electric Utilities.

41 Cybersecurity and Infrastructure Security Agency (CISA), People's Republic of China State-Sponsored Cyber Actor Living off the Land to Evade Detection, May 24, 2023, https://www.cisa.gov/news-events/cybersecurity-advisories/aa23-144a

well. As with natural hazards, the same attacks that produce such outages may also damage or disrupt restoration infrastructure. Moreover, precisely because blackstart power restoration is so crucial for societal continuity and national security, attackers may prioritize the disruption of blackstart infrastructure to help achieve their goals in the conflict and continue attacking the restoration process until they prevail or are defeated.

Such attacks could extend the already lengthy time (likely a week or more) required to complete the restoration process from the initial energizing of blackstart units to the reestablishment of major interconnections. For water systems, hospitals, and other facilities that will rely on emergency power during blackouts, such extended timelines would endanger and ultimately halt their ability to continue delivering vital services. Furthermore, given the risk that adversaries will target and sustain cyberattacks against blackstart infrastructure and operations, such manmade threats will entail resilience challenges over and above those posed by natural hazards.

2. The Changing Resource Mix

The transformation of the electric grid poses an additional challenge to blackstart resilience. Driven in part by Federal, state, and local goals to decarbonize power production, the electric systems in the U.S. and many partner nations are undergoing a revolution in their generation fleets. The shift from fossil fuel-based generation (especially coal) to solar and wind power is creating problems for maintaining the operating reliability of the bulk power system, and also for ensuring the resource adequacy of generation for blackstart. For many decades, natural gas-fired generators (together with hydropower, where available) have been crucial for conducting restoration operations. Regulatory and statutory requirements for decarbonization will soon put the availability of such resources at risk, and magnify concerns over resource adequacy for blackstart operations that Regional Transmission Organizations (RTOs) and other grid operators already face.

Over the longer term, the inverters that help control power flows from solar arrays, wind turbines, and battery energy storage systems may become capable of supporting blackstart restoration, and supplement – or even replace – the role played by gas generators. Supporting advances in the capabilities of these inverter based resources (IBRs) is essential. At present, however, IBRs pose a threat to grid reliability, due in part to their tendency to trip offline when grid disturbances occur. The Handbook examines the technical challenges and opportunities to mitigate these risks and thereby enable IBRs to contribute to blackstart, especially through the development of grid-forming (GFM) inverters that can help restart the electric system when outages occur. For the foreseeable future, however, specialized regulatory and other measures will be necessary to sustain the contributions of gas generators to restoration.

3. Electrification, Cross-Sector Interdependencies, and Societal Continuity

The National Academy of Engineering hosted a workshop in October 2022: "Creating A Sustainable National Electric Infrastructure While Maintaining Reliability and Resiliency of the Grid".[42] This report indicated the need for a sectoral or cross-sectoral definition of resilience to enhance infrastructure systems to operate reliably under normal conditions and adapt, withstand, and more rapidly recover from a growing number of high-impact low-frequency events. Recommendations indicated the necessity for integrated resource and T&D planning and investment prioritization, the creation of grid resilience standards that explicitly address failures that cascade across sectoral interdependencies, and the need for probabilistic assessments to account for "long tail" events on power grids.

After years of minimal growth in the demand for power in many nations (and even some recent periods of decline), the electrification of multiple infrastructure sectors will create new systemic dependencies on the grid. Requirements to serve data centers, hydrogen production, and other new types of loads will further diversify the demand for resilient power. Moreover, the expansion of these electricity-dependent systems and critical functions will exacerbate the risk of cross-sector failures when outages occur, and greatly complicate blackstart plans and capabilities to ensure societal continuity.

The operation of compressor stations and other electric-powered natural gas infrastructure offers a prime example of the imperative to assess and

---

42 National Academy of Engineering, Creating A Sustainable National Electric Infrastructure While Maintaining Reliability and Resiliency of the Grid. February 2023. Report available at https://www.nae.edu/289966/Public-Documents

mitigate the risks to blackstart posed by cascading failures between interdependent sectors. As more and more sectors are electrified, these interdependencies will grow increasingly tight and complex. Building heating, transportation infrastructure, critical manufacturing, and myriad other electrified assets and functions (and the supply chains that support them) will present new challenges for understanding the interactions and societal effects of wide-area power outages, and for emergency management operations to save and sustain lives.

Accounting for these risks in blackstart planning – including prioritizing the restoration of power among the many competing loads vital to public health and safety – will also become increasingly difficult and politically fraught. Transmission operators and their partners have detailed restoration plans to prioritize restoration to nuclear power plants, gas infrastructure, hospitals, and other loads essential for power restoration and public health and safety. The same is true for restoring service to Critical Defense Facilities identified by the Secretary of Energy as required by Section 215A of the Federal Power Act.[43] With the electrification of so many other sectors essential for public health and safety, national security, and the economy, prioritizing the restoration of service between these critical customers (and doing so in ways that account for potential damage to blackstart infrastructure) will pose immense, crucial challenges.

To meet the needs of the affected population – indeed, to support population continuity during the blackstart restoration process – the preplanned architecture and priorities of that process must be designed to ensure the earliest possible, timely restoration of the lifeline infrastructures and other critical facilities that communities, and the nation as a whole, require.

In summary, the urgency of robust, highly effective blackstart processes – including the supplemental tools and capabilities needed due to the society-wide impacts of a long term outage – cannot be exaggerated. And in any review

43 16 USC 824o-1: Critical electric infrastructure security, https://uscode.house.gov/view.xhtml?req=(title:16%20section:824o-1%20edition:prelim)

of blackstart requirements, it is particularly urgent to recognize that the purpose of blackstart is not simply to reestablish grid operations. To meet the needs of the affected population – indeed, to support population continuity during the blackstart restoration process – the preplanned architecture and priorities of that process must be designed to ensure the earliest possible, timely restoration of the lifeline infrastructures and other critical facilities that communities, and the nation as a whole, require.

4. Structure of Chapter I and the Handbook as a Whole

Section B of this chapter provides a primer on blackstart restoration and the phases it encompasses. Section C offers an overview of the principal regulatory standards that apply to blackstart and support its effectiveness. Section D offers recommendations on how to build on these standards, especially for strengthening blackstart training and exercises. Section E describes how the Handbook is structured to propose additional regulatory initiatives, policy options, and resilience strategies to address natural and manmade Black Sky hazards.

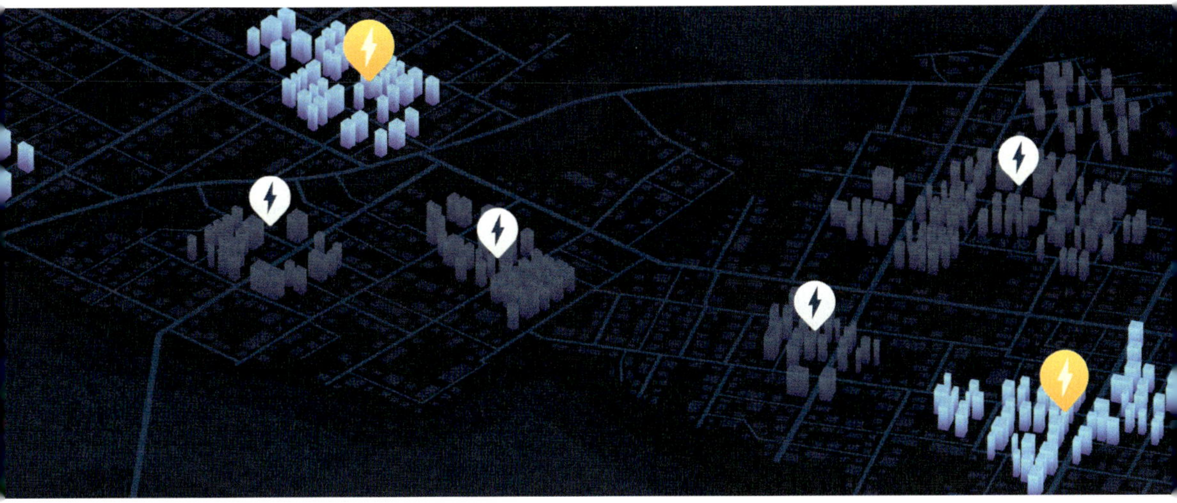

I | BLACKSTART 101

Background

In most nations, transmission operators (TOPs) and their electricity subsector partners have extensive plans and capabilities for blackstart. In the continental United States, Canada, and the northern portion of Baja California, Mexico, the North American Electric Reliability Corporation (NERC) reinforces these capabilities with mandatory standards for blackstart restoration. NERC's Critical Infrastructure Protection (CIP) standards for the bulk power system also provide broader cyber, physical, and supply chain security requirements for high voltage transmission systems and other assets on which blackstart depends. Regional Transmission Organizations (RTOs), Independent System Operators (ISOs), and other BES entities help ensure the availability of adequate blackstart resources in organized markets. In addition, the U.S. Department of Energy (DOE), which is the Sector Risk Management Agency (SMRA) for the energy sector in the United States, assists TOPs and their partners in building resilience against intensifying threats to blackstart infrastructure and operations. An extensive international literature has also emerged on technical aspects of blackstart.[44]

In society as a whole, however, the realm of blackstart power restoration is largely unknown. That lack of awareness is understandable, particularly in the United States. While blackstart operations have been needed on multiple occasions in Hawaii and Puerto Rico, in the continental United States in response to all previous blackouts, including those caused by major hurricanes, earthquakes, and other natural hazards, utilities have restored power primarily by bringing electricity into a blacked-out region from the unaffected areas surrounding it. TOPs and their partners then use that externally provided power to restart generators and restore service to hospitals, nuclear power plants, and other critical customers.

These traditional "outside-in" restoration operations benefit from the intensive, sustained measures that utilities and their regulators take to prevent local grid disruptions from cascading into surrounding regions. As long as the electric systems in surrounding regions of the same power grid remain operational, they can send power to re-energize the grid in stricken areas through efficient and long-established procedures, though Superstorm Uri and other severe, multi-state weather events have recently stressed the ability of utilities to implement such traditional restoration strategies.

Black Sky hazards, however, can create grid disruptions vastly larger than in winter storm Uri (2021) or any other previous U.S. weather events. Comprehensive, sophisticated attacks by China or other potential adversaries risk creating multi-region outages where it will be essential to restart the grid from within the blacked out area, while also (wherever possible) importing power from any contiguous areas where high voltage systems are still functioning. Magnifying these risks: both the complexity and interconnectivity across infrastructure sectors are growing so rapidly that the resulting, organically integrated, tightly interdependent systems are becoming increasingly fragile,

44 H. Khoshkhoo, et al, "Survey of Power System Restoration Documents Issued from 2016 to 2021," International Transactions on Electric Energy Systems, Wiley/Hindawi, October 2022, https://www.hindawi.com/journals/itees/2022/1754013/

and are systematically lowering the threshold for what scale of events can create catastrophic outages and necessitate blackstart. Those same Black Sky events will likely disrupt the infrastructure necessary for both inside-out and outside-in restoration. Moreover, once initial blackouts have occurred, adversaries may seek to disrupt restoration operations to extend the duration and life-threatening consequences of such outages.

Survivable microgrids and other emergency power options can help stave off threats to public safety and national critical functions, which the Cybersecurity and Infrastructure Security Agency (CISA) defines as "functions of government and the private sector so vital to the United States that their disruption, corruption, or dysfunction would have a debilitating effect on security, national economic security, national public health or safety, or any combination thereof."[45] A growing number of Level-1 trauma centers, military bases, major water systems, and other crucial assets are strengthening their ability to sustain their critical functions if cyberattacks cut off their grid-provided power.

Survivable microgrids and other emergency power options can help stave off threats to public safety and ensure national critical functions.

At present, however, extended outages will eclipse the ability of most facilities to sustain such emergency operations for more than a few days. These limitations are especially significant given the multi-week timelines that would likely be required to restore functioning interconnections via the blackstart process, especially given the damage that blackstart-related infrastructure could have incurred in the precipitating event (and, potentially, in follow-on adversary attacks). Moreover, especially for non-Defense facilities, their emergency power systems may themselves be disrupted by Black Sky hazards. The net effect: for NCFs and society as a whole, the effective and carefully prioritized restoration of grid-provided power will be essential to save and sustain lives and prevent cascading failures across multiple, interdependent infrastructure sectors.

45 Cybersecurity and Infrastructure security Agency (CISA), National Critical Functions, https://www.cisa.gov/national-critical-functions

Blackstart plans and capabilities can make that restoration possible. Rather than rely on external power to restart generators that have tripped offline in a Black Sky event, blackstart resources can do so autonomously. According to the FERC-NERC-Regional Entity Joint Review of Restoration and Recover Plans, which offers an exceptionally useful analysis of blackstart restoration, a blackstart resource is defined as:

> *A generating unit(s) and its associated set of equipment which has the ability to be started without support from the System or is designed to remain energized without connection to the remainder of the System, with the ability to energize a bus, meeting the Transmission Operator's restoration plan needs for Real and Reactive Power capability, frequency and voltage control, and that has been included in the Transmission Operator's restoration plan.*[46]

The analysis that follows provides an overview of how these blackstart restoration plans function in sequential phases. Executing the multi-phase process required for blackstart will entail major challenges, especially if 1) the Black Sky event necessitating restoration damages blackstart-related infrastructure, and 2) adversaries sustain their attacks to disrupt and delay the restoration process. Each of these phases also entails opportunities to strengthen the resilience of blackstart plans and capabilities. The analysis that follows provides an overview of the steps that blackstarting the grid will entail.

Initial Steps in the Blackstart Process

Each TOP has an individualized blackstart restoration plan tailored to their respective service areas. These plans reflect the topography of the generation, transmission, and distribution assets that system operators will employ, as well the location and attributes of the loads they must pick up as restoration goes forward, including 1) critical loads that enable restoration, including substations

46 NERC, Report on the FERC-NERC- Regional Entity Joint Review, p. 113

and electric-powered natural gas compression stations; and 2) priority loads that are essential for societal continuity and national security, including water systems, hospitals, and major defense installations.[47] Restoration of auxiliary power to nuclear power plants is also a key priority for TOPs that have such plants in their service areas.

In addition to the differing load pickup requirements that TOPs will have, these operators also differ in the approach they take to blackstart restoration. In particular, while systems typically follow a "bottom up" approach that entails the formation of power islands within a blacked out area, significant variations in their restoration procedures exist (i.e., core island, backbone island, etc.).[48] Nevertheless, despite these system-specific features and differences in terminology used by transmission operators, most blackstart restoration plans entail four basic phases.

> If EMS and SCADA systems are disrupted or not reliable, operators may need to employ "spare tire" capabilities and operations as a fallback.

1. Establishing Situational Awareness and (as needed) Adjustment of Restoration Plans

In the aftermath of the event that requires blackstart, TOPs will first assess the conditions of the system, including: 1) the availability of transmission and generation resources necessary to execute the TOP's blackstart plan; 2) communications capabilities; and 3) energy management systems (EMS) and supervisory control and data acquisition (SCADA) system status. As will be discussed later in the Handbook, gathering data on potential disruptions to natural gas transmission systems could be enormously helpful in Black Sky events.

47 Restoration of auxiliary power to nuclear power plants is also a key priority for TOPs that have such plants in their service areas.

48 For an especially useful and detailed description of the typical steps entailed in power restoration, and variations in TOP approaches to blackstart restoration, see PJM, Service Restoration, slides 67-105. https://www.pjm.com/-/media/training/nerc-certifications/trans-exam-materials-2020/transmission-itp/lesson-10-system-restoration.ashx

Based on these assessments, operators will then determine whether their primary restoration plan remains viable, or whether they need to employ and/or modify their backup restoration plans.[49] Operators will also need to ensure that they can communicate with their personnel in substations along the restoration path, as well as with 1) generating stations; 2) reliability coordinators and balancing authorities; 3) neighboring TOPs; 4) law enforcement agencies, emergency managers, and other relevant government officials; and 5) other key participants in the overall restoration process. In addition, if EMS and SCADA systems are disrupted or not reliable, operators may need to employ "spare tire" capabilities and operations as a fallback.[50]

These initial steps may prove enormously difficult to execute in the wide area blackouts that characterize Black Sky events. The Electric Reliability Council of Texas (ERCOT) notes that "pre-established plans and procedures cannot foresee all the possible combinations of system problems that may occur after a major failure."[51] Ensuring the availability of resilient and redundant communications systems will be a prerequisite for assessing these complex problems and adjusting restoration plans accordingly. For some hazards, including EMP attacks, the replacement of damaged equipment with (securely stored) spares may be necessary to facilitate the execution of either primary or backup restoration plans. Against cyberattacks, assessing whether malware is still embedded in critical networks or restoration infrastructure (protective relays, etc.) will be essential as well. Otherwise, once restoration is underway, that malware may cause disruptions that cascade back along the restoration path and require the re-start of the blackstart process.

A number of initiatives are underway to help TOPs and their partners improve their ability to assess the state of their systems and degree of damage

49 CIGRE, Power System Restoration Accounting for a Rapidly Changing Power System and Generation Mix, August 2023, 13, https://www.cigre.org/article/GB/power-system-restoration-accounting-for-a-rapidly-changing-power-system-and-generation-mix

50 North American Transmission Forum, Bulk Electric Systems Operations Absent Energy Management System and Supervisory Control and Data Acquisition Capabilities—a Spare Tire Approach, 2024, https://www.natf.net/docs/natf/documents/resources/system-operations/bes-operations-absent-ems-and-scada-capabilities---a-spare-tire-approach---open.pdf

51 Electric Reliability Council of Texas (ERCOT), ERCOT Nodal Operating Guides Section 4: Emergency Operation, June 1, 2017, p. 4-16, http://www.ercot.com/content/wcm/current_guides/53525/04-060117.doc.

to their restoration infrastructure. For example, image processing methods are under development to analyze satellite imagery obtained before and after hurricane-induced blackouts, with Artificial Intelligence (AI) tools employed to use that data to assess damage to electric system infrastructure.[52] Operators can also create plans to optimize the restoration by connectivity to Phasor Measurement Units (PMUs) and their data on grid conditions to facilitate restoration in highly disrupted environments.[53] Continued support for such initiatives will be essential to improve system assessment capabilities for Black Sky events.

2. Starting Up the Blackstart Resource and Energizing the Cranking Path

Once restoration plans are revised as needed and supporting communications links are up and running, operators will start up one or more blackstart resources to initiate the broader restoration process. Once the blackstart resource is operating, the power it produces is transmitted along a dedicated "cranking path" to sequentially energize substations and additional generators. That latter function reflects the fact that most generators require grid-provided power before they can restart. When wide-area electric outages occur and generators go offline within the blacked out region, those generators require power to return to operation. Power is essential for the auxiliary equipment that starts a generating unit and keeps it running (e.g., oil pumps, gas compressors, draft fans, controls, switchgear, and sensors, and safety and protection systems). Generation plants also require power for "house loads," such as lights, data systems, and telecommunications equipment.[54] The cranking path delivers this power.

52 R. Jamshad, M. U. Qureshi, and S. Grijalva, "Geographic information systems (GIS) image analysis for prioritizing power system restoration," Clemson University Power Systems Conference (PSC), Clemson University, Charleston, SC, USA, 2018, https://ieeexplore.ieee.org/abstract/document/8664080

53 S. N. Edib, Y. Lin, V. M. Vokkarane, F. Qiu, R. Yao, and D. Zhao, "Optimal PMU restoration for power system observability recovery after massive attacks," IEEE Transactions on Smart Grid, vol. 12, no. 2, pp. 1565–1576, 2021, https://ieeexplore.ieee.org/document/9215000

54 Jose R. Garcia, et al, Hydropower Plants as Black Start Resources, Oak Ridge National Laboratory, May 9, 2019, 7, https://www.energy.gov/sites/prod/files/2019/05/f62/Hydro-Black-Start_May2019.pdf

NERC defines a cranking path as:

> *"a portion of the electric system that can be isolated and then energized to deliver electric power from a generation source to enable the startup of one or more other generating units."*[55]

The initial generation unit is typically a hydroelectric, gas turbine, or diesel generator (with stored on-site fuel) that can be started with local battery power to begin restoration along the cranking path.[56] The power produced by that blackstart unit then re-energizes the next generator and associated infrastructure (substations, etc.) on that pre-designated cranking path. As new generation assets are brought online, system operators pick up load to balance and stabilize that additional generation and strengthen the stability of the small but growing power island that the cranking path has established, ideally picking up critical loads where possible.[57]

3. Next Start for Island Expansion

Bringing larger and more numerous generators online not only enables grid operators to service more priority loads as island growth goes forward, but also enhances the stability of the island by reducing the risk that the loss of any single generator will result in island collapse, and also by giving operators additional resources to manage frequency and voltage excursions.

---

55 NERC, Glossary of Terms, p. 10.

56 These are often small units that cannot themselves restore the majority of loads in a service area. Instead, they power larger nearby generators and associated service station loads required to begin large-scale restoration, and may also provide power to other systems critical to the restoration process such as communications and fuels for power generation. See: Carlos Grande-Moran and James W. Feltes (Siemens Power Technologies International Inc.), "An Overview of Restoration Issues and Blackstart Analysis," paper presented at the Minnesota Power Systems Conference 2014, Minneapolis, MN, November 4, 2014, p. 2 and 4, http://cce.umn.edu/documents/CPE-Conferences/MIPSYCON-Papers/2014/AnOverviewofRestorationIssuesandBlackStartAnalysis.pdf.

57 An electric island is "An electrically isolated portion of an Interconnection. The frequency in an electrical island must be maintained by balancing generation and load in order to sustain operation. Islands are frequently formed after major disturbances wherein multiple transmission lines trip, or during restoration following a major disturbance." FERC-NERC-Regional Entity Report, p. 114

Next start generators are critical to this growth process. A seminal FERC/NERC study on blackstart generation, Blackstart Resource Adequacy (May 2018) defined "next start" generating units as:

"the first generating unit in the cranking path to be energized using power from the blackstart generating unit."[58]

However, in more recent usage, that definition has broadened to include the many additional generators that will be required to grow power islands both within and beyond the initial cranking path.

Many system operators will seek to expand the initial islands they create to reach a defined level of inertia that would allow them to handle a specific, predetermined number of generation outage contingencies (i.e., N-X) prior to synchronization with neighboring islands. But maintaining the stability of these islands will be both vital and fraught with difficulties. For example, as operators pick up loads to match the generation they are adding, cold load pickup can result in voltage drops that can re-collapse the system.[59]

Failure to maintain balance between generation and load can also cause equipment damage, island collapse, and force operators to reinitiate the entire blackstart process. In traditional outside-in restoration, the large portions of the grid that remain intact have much greater capacity to absorb changes in load without frequency deviations. However, smaller blackstart islands are much more volatile and difficult to balance. Blackstart operators must add loads in balance with additional generation, while maintaining sufficient stability of the system's frequency and voltage, to prevent re-energized generators from tripping off. The need to energize loads that have been offline for an extended period of time will make it all the more difficult to conduct balancing operations.[60]

58 FERC/NERC,, Blackstart Resource Adequacy, May 2018, 59, https://www.ferc.gov/sites/default/files/2020-05/bsr-report.pdf

59 Y. Jiang, S. Chen, C.-C. Liu et al., "Blackstart capability planning for power system restoration," International Journal of Electrical Power & Energy Systems, vol. 86, pp. 127–137, 2017, https://www.sciencedirect.com/science/article/abs/pii/S0142061516307220?via%3Dihub

60 This type of restoration involves cold load pickup, which is bigger and will last a while then taper off, complicating power management. DOE, Hydropower Plants as Black Start Resources, pp. 2, 9. For associated concerns regarding inrush currents for cold load pickup, see p. 2.8 and Grande-Moran and Feltes (Siemens), "An Overview of Restoration Issues and Blackstart Analysis," p. 9-10.

These challenges highlight broader implications for blackstart resilience. Even under the best of circumstances, conducting the initial phase of blackstart restoration and maintaining the initial power island's stability poses significant difficulties. Yet a Black Sky event which could precipitate the need for blackstart will constitute the worst of circumstances. The same extreme event that caused the wide area blackout will likely have disrupted the dedicated transmission lines, substations and other assets that help comprise the cranking path, unless these assets have been adequately hardened against Black Sky effects. The same is true of control centers and generators along that path, as well as the communications systems and flows of natural gas necessary to execute many TOP plans.

At the same time that operators are adding generation, they also control the pickup of load to keep generation and load balanced and to provide for frequency and voltage control within the emerging power island that the blackstart restoration process creates. The cranking path and its transmission lines are initially isolated from surrounding transmission and distribution infrastructure; energizing the cranking path creates an island of power within the blacked out region. Operators then grow that island by adding generation and picking up new load through a carefully controlled, complex process that maintains the island's stability. Subsequent portions of the Handbook examine the operational difficulties inherent in this process of island growth. For this overview of the blackstart process, however, suffice it to say that the difficulties of managing the growth and stability of such islands will be all the greater in a disrupted Black Sky environment.

> Assessments of the adequacy and resilience of blackstart generation assets must include both the blackstart resources that initiate the restoration process, and next start generators essential for carrying that process forward.

A number of BES entities will play critical roles in this initial phase, including local utilities, TOPs, Transmission Owners (TOs), Regional Transmission Operators (RTOs), Independent System Operators (ISOs), and Reliability Coordinators (RCs), and Generator Operators (GOPs). These entities will attempt to "communicate with one another to ascertain the condition of the grid, the precipitating event(s), and the extent of the damage."[61] These assessments will also include an examination of the blackout's impact on priority loads, transmission/distribution paths, and interdependent infrastructure systems.[62]

Their ability to share and jointly develop these assessments will be at risk until they have communications networks (and the power those networks need) to do so. Survivable communications will also be necessary for coordination with state and local authorities and emergency managers responsible for assessing and mitigating the outage's consequences. In addition, TOPs and their partners will need to continually assess the status of infrastructure along the cranking path, the readiness of critical customers (including military bases) to reconnect with the grid and accept power, and other factors.

Assessments of the adequacy and resilience of blackstart generation assets must include both the blackstart resources that initiate the restoration process, and next start generators essential for carrying that process forward. Moreover, in many regions of the U.S. and partner nations, gas-fired generators provide the vast majority of next start assets. Federal and state decarbonization goals and regulations will put these assets at growing risk, and create new challenges for ensuring adequate blackstart resource availability – and new incentives to explore the use of IBRs to supplement such fossil generation.

4. Synchronizing Power Islands

As TOPs continue to grow and maintain the stability of their electric islands, the next phase of blackstart will be to synchronize and integrate these islands and, ultimately, reestablish operation of the interconnections that had been blacked out. For the United States, in an extreme Black Sky event, this could mean

61 DOE, Hydropower Plants as Black Start Resources, p. 2.4. ERCOT emphasizes the importance of establishing communications immediately following an outage that requires blackstart, but also notes that doing so might not be possible. If TOPs cannot establish communications with ERCOT, they are to "evaluate system conditions and proceed independently" with their blackstart plan. See: ERCOT, ERCOT Nodal Operating Guides Section 4: Emergency Operation, p. 4-17; ERCOT, ERCOT Nodal Operating Guides Section 8 Attachment A: Detailed Black Start Information, November 2016, p. 8A-1. Similarly, the first step of PJM's high-level restoration strategy is to "Perform a system assessment to determine extent of outage." See: PJM, Manual 36: System Restoration (Revision 27), effective December 3, 2019, p. 12, 18, and 37, https://www.pjm.com/~/media/documents/manuals/m36.ashx.

62 DOE, Hydropower Plants as Black Start Resources, p. 2.5.

restarting operations of all three of the interconnections that serve the nation.

In a multi-region blackout, many TOPs will be simultaneously conducting blackstart operations. A given entity may also be constructing multiple cranking paths and initial islands within its own service area to expedite restoration. Once two adjacent islands are sufficiently stable (i.e., enough load and generation to handle minor contingencies), operators must synchronize the islands by matching voltages, phase angles, and frequencies at the connection points to consolidate them.[63] TOPs and their partners will repeat this process until the entire interconnection is restored.

Discussions with TOP personnel suggest that inter-island load and generation sharing is a challenging process that has not been tested or demonstrated effectively.

This step will be technically complex, as each individual island will likely be out of synchronism with surrounding islands. To connect two islands, operators will need to coordinate their respective frequencies to achieve synchronization.[64] Failure to do so prior to connecting their islands could damage equipment and cause generators to trip off, forcing the TOPs to go back to the beginning of the restoration process. The tasks involved will be even more difficult if sensing and automation systems necessary to support synchronization lack the power they need – or worse, are disrupted due to the precipitating Black Sky event, whether natural or adversary-induced. In addition, discussions with TOP personnel suggest that inter-island load and generation sharing is a challenging process that has not been tested or demonstrated effectively.

Once an RC has integrated all the islands in their own reliability control area, they must synchronize it with neighboring RCs to reconstitute the grid. An RC may also decide to connect to neighboring RCs before all islands in the RC control area are consolidated, based on resource availability and other factors. In either case, NERC standards require RC restoration plans to include

63 DOE, Hydropower Plants as Black Start Resources, p. 2.12. RTOs/ISOs overseeing blackstart operations might have "designated synchronization points" for connecting islands. See: ERCOT, ERCOT Nodal Operating Guides Section 4: Emergency Operation, p. 4-18.

64 This is conceptually similar to synchronizing a generator to the system on startup but will be more difficult because of the islands' lower system inertia.

criteria, conditions, and training programs for re-establishing interconnections with other RCs.[65] RCs will face similar challenges when connecting RC areas as TOPs do when connecting individual islands – each RC area will likely be out of synchronism with neighboring areas. RC areas have more inertia and will therefore be easier to keep stable. Nevertheless, closure and matching synchronism is still critical. RCs will only have completed their NERC-mandated blackstart duties when all of its TOPs are interconnected, and each RC area is connected to neighboring areas.[66]

Moreover, existing equipment in some BES systems may lack adequate control interfaces necessary to facilitate load sharing between islands. The Department of Energy should consider sponsoring additional research and development by the electric industry and National Laboratories to clarify these challenges and develop new options to overcome them.

> The time required for completion will vary with multiple factors, but most transmission system operators, reliability coordinators, and their partners anticipate that multiple weeks could be necessary.

As blackstart restoration proceeds, entities will ultimately need to transition load demand and generation resource balancing to their respective Balancing Authorities (BAs). NERC standards for TOPs and RCs identify criteria and operating processes for "transferring operations [and authority] back to the Balancing Authority" as necessary components of system restoration plans.[67] However, The FERC-NERC-Regional Entity report on blackstart restoration notes that uncertainties remain as to how this transfer will occur. The report found that "while Requirement R1.9 requires the transmission operator to have processes for transferring authority back to the balancing authority in accordance with the reliability coordinator's criteria, the joint staff review team found that some restoration plans contain

65 NERC, EOP-006-3 – System Restoration Coordination, effective April 1, 2019, R1.2 and R7, https://www.nerc.com/pa/Stand/Reliability%20Standards/EOP-006-3.pdf.

66 NERC, EOP-006-3, R1.

67 NERC, EOP-005-3 – System Restoration from Blackstart Resources, effective April 1, 2019, R1.9, https://www.nerc.com/pa/Stand/Reliability%20Standards/EOP-005-3.pdf; NERC, EOP-006-3, R1.6.

limited information on the triggers, steps involved, or checks necessary for that transfer."[68] That said, BAs are not likely to regain their full operational control and coordination functions until the grid is operating within stable parameters. Blackstart operators will need to restore power to critical defense installations long before that status is achieved.

Finally, entities will need to continuously monitor and maintain stability and reliability as blackstart operations go forward. While all entities will be working to ensure cranking paths and islands remain intact, RTOs/ISOs (or RCs in some service areas) play a particularly important role in overseeing grid conditions, reinforcing the need for survivable communications systems to establish shared, wide-area situational awareness.[69] These tasks will not be "complete" until the interconnection is restored, and these entities transition back to normal, day-to-day operations.[70] The time required for completion will vary with multiple factors, but most transmission system operators, reliability coordinators, and their partners anticipate that multiple weeks could be necessary. Those timelines would be further extended if essential resources are unavailable to support the latter phases of the restoration process, especially in terms of the flow of gas to natural gas-fueled next start generators, emergency communications systems, or other interdependent infrastructure. Moreover, the phases outlined above will not be distinct or well-defined. Entities involved in the FERC-NERC review noted that:

> *"Estimating the time to complete a restoration plan and fully restore the system is a difficult task. Many factors must be taken into consideration, including the extent of the blackout, damage to the system, state of generating units and the system status of neighboring utilities. In the future this will include challenges with inverter-based resources (IBR) at both the transmission and distribution system levels."*

68 FERC -NERC, Review of Restoration and Recovery Plans (2016), p. 16.

69 See, for example, ERCOT, ERCOT Nodal Operating Guides Section 8 Attachment A: Detailed Black Start Information, November 2016, p. 8A-6.

70 Following a black sky event, initial day-to-day operations may not be fully normal, especially if equipment is damaged. For example, if multiple transmission or distribution transformers are damaged, the TOP and distribution system operators may have to implement rotating blackouts, even though the interconnection is back in operation.

These IBR-related challenges are critically important. In fact, it is crucial to keep in mind that, if blackstart operations are not configured to quickly support lifeline critical infrastructure facilities, the survival of affected communities may be at risk, and the consequent scale of the catastrophe could become unimaginable. Before exploring them, however, it will first be helpful to analyze the mandatory standards that support blackstart effectiveness and provide a foundation for additional resilience measures in the future.

II | FERC, NERC, AND CURRENT REGULATORY FOUNDATIONS OF BLACKSTART RESILIENCE IN NORTH AMERICA

Systems for regulating power grids vary widely between the nations in North America and those in Europe, Asia, and other parts of the globe. Nevertheless, the North American model highlights regulatory challenges and options to meet them that are widely shared by other regions that are undergoing drastic changes in their generation fleets as well.

FERC is the federal regulator of interstate transmission of electricity with responsibility to review and approve standards that provide for the reliable operation of the bulk power system. In addition, FERC oversees NERC, which is the federally designated U.S. electric reliability organization (ERO). NERC is responsible for conducting reliability assessments and enforcing mandatory standards to ensure the reliability of the bulk electric system or BES—a term that refers to (1) facilities and control systems necessary for operating the electric transmission network and (2) the output from certain generation facilities needed for reliability. NERC develops reliability standards collaboratively with the electric industry and then sends them to FERC, which can either approve them or remand them to NERC for revision.[71]

These reliability standards include critical infrastructure protection standards for protecting electric utility-critical and cyber-critical assets from cyberattacks. These Critical Infrastructure Protection (CIP) standards are discussed in Chapter VIII; they provide important elements of a regulatory foundation for efforts to improve overall grid resilience. In addition, NERC has established specialized EOP standards for blackstart training, simulations, and related activities.

Blackstart Standards

EOP-005-3 – System Restoration from Blackstart Resources requires TOPs to maintain restoration plans that "shall be implemented to restore the Transmission Operator's System following a Disturbance in which one or more areas of the Bulk Electric System (BES) shuts down and the use of Blackstart Resources is required to restore the shutdown area…"[72] To ensure the adequacy of blackstart plans and capabilities, EOP-005-3 also provides requirements for blackstart testing by BES entities.

Under **R6** of that standard, each TOP must "verify through analysis of actual events, a combination of steady state and dynamic simulations, or testing that its

71 U.S. Government Accountability Office, Critical Infrastructure Protection: Actions Needed to Address Significant Cybersecurity Risks Facing the Electric Grid, August 29, pp. 13-4, https://www.gao.gov/assets/gao-19-332.pdf Note that NERC reliability standards and FERC regulation apply to the Electric Reliability Corporation of Texas (ERCOT) even though ERCOT is considered intra-state for other regulatory purposes.

72 North American Electric Reliability Corporation (NERC), EOP-005-3 – System Restoration from Blackstart Resources, effective April 1, 2019, R6, https://www.nerc.com/pa/Stand/Reliability%20Standards/EOP-005-3.pdf.

restoration plan accomplishes its intended function."[73] This analysis, simulations or testing (which must be conducted at least every five years) must verify:

6.1. The capability of Blackstart Resources to meet the Real and Reactive Power requirements of the Cranking Paths and the dynamic capability to supply initial Loads.

6.2. The location and magnitude of Loads required to control voltages and frequency within acceptable operating limits.

6.3. The capability of generating resources required to control voltages and frequency within acceptable operating limits.

R7 provides additional testing requirements for TOPs to "verify that each Blackstart Resource is capable of meeting the requirements of its restoration plan."[74] At least once every three years, TOs must conduct tests of:

7.2.1. The ability to start the unit when isolated with no support from the BES or when designed to remain energized without connection to the remainder of the System.

7.2.2. The ability to energize a bus. If it is not possible to energize a bus during the test, the testing entity must affirm that the unit has the capability to energize a bus such as verifying that the breaker close coil relay can be energized with the voltage and frequency monitor controls disconnected from the synchronizing circuits.

R8 requires TOPs to conduct annual system restoration training as part of its operations training program, including training on:

8.1. System restoration plan including coordination with its Reliability Coordinator and Generator Operators included in the restoration plan.

8.2. Restoration priorities.

8.3. Building of cranking paths.

8.4. Synchronizing re-energized sections of the System.

73 NERC, EOP-005-3, R6.

74 NERC, EOP-005-3, R7.

8.5. Transition of Demand and resource balance within its area to the Balancing Authority.[75]

NERC also has an RC-specific standard for restoration plans, which includes requirements for developing and implementing blackstart restoration plans, reviewing the plans of TOPs in their RC area, and blackstart-specific training requirements.[76] These planning and testing requirements form the regulatory basis of current blackstart preparedness for TOPs and other BES entities subject to NERC standards.

75 NERC, EOP-005-3, R7.

76 NERC, EOP-006-3 — System Restoration Coordination, R1, R5, and R7, https://www.nerc.com/pa/Stand/Reliability%20Standards/EOP-006-3.pdf.

III | BUILDING ON THE REGULATORY FOUNDATIONS OF BLACKSTART: STRUCTURE OF THE HANDBOOK

NERC's mandatory standards, regulatory guidelines, and ongoing research initiatives provide an absolutely vital framework for strengthening blackstart preparedness. This Handbook offers recommendations for building on that framework to address specific Black Sky restoration challenges, including those posed by the increasing penetration of IBRs into the grid.

Before describing the Handbook's structure, however, it will first be useful to address an overarching, high-impact opportunity: the development of additional training and exercise initiatives to bolster the resilience of today's blackstart systems and those that will evolve to meet future needs.

A. Training, Simulating and Exercising Blackstart for Real-World Blackstart Conditions

The plans, testing, training, and simulations required by NERC's standards are not sufficient to ensure that TOPs and their partners will be prepared to conduct blackstart restoration in the highly disrupted situations that will follow extreme, Black Sky events, which represent likely scenarios for large scale outages that will require blackstart.

One omission is particularly problematic: EOP-005-3 provides no guidance on how tests ought to account for the other challenges that cyberattacks might create for blackstart operations. BES entities must already comply with NERC's Critical Infrastructure Protection (CIP) cybersecurity standards; by implication, blackstart assets that fall within those infrastructure mandates would need to be protected accordingly.[77] But no specific guidelines or requirements help BES entities ensure that their tests reflect the environment in which they would have to execute their plans, other than that they would have no BES-provided power with which to do so.

In any large-scale cyberattack on the power grid by China or other adversaries, we should expect that strikes will also occur against the communications systems on which transmission operators, RCs, BAs, Generator Owners (GOs), and other key participants in the blackstart restoration process will rely.

In particular, given the many weeks (or longer) timelines associated with any realistic blackstart scenario, effective, realistic blackstart preparations should be structured to account for the complex, highly disrupted environment in which real-world restoration operations would go forward. Key planning factors to incorporate in training, simulations and exercises include:

77 See, for example: North American Electric Reliability Corporation (NERC), CIP-002-5.1a – Cyber Security — BES Cyber System Categorization, December 27, 2016; NERC, CIP-003-6 – Cyber Security — Security Management Controls, July 1, 2016; NERC, CIP-007-6 – Cyber Security — System Security Management, July 1, 2016.

1. Disrupted communications.

In any large-scale cyberattack on the power grid by China or other adversaries, we should expect that strikes will also occur against the communications systems on which transmission operators, RCs, BAs, Generator Owners (GOs), and other key participants in the blackstart restoration process will rely. NERC's November 2022 GridEx VI highlighted the imperative to "strengthen operational coordination between the electricity industry and communications providers." The GridEx VI Tabletop scenario featured a widespread loss of landline and cellular communications while electricity utilities were recovering from the cyber and physical attacks and restoring the grid. Exercise participants "agreed that the loss of communications would essentially halt the grid restoration process."[78]

A growing number of TOPs and other BES entities are accounting for the risk of cyber-or EMP-induced disruptions in their communications links. As will be discussed below, some of these entities are also training their personnel to help establish and/or operate secure emergency communications networks. These measures constitute best practices and should be adopted across the BES – alongside collaboration with communications providers – to develop and deploy systems that can survive increasingly sophisticated Chinese attack capabilities (including satellite-based networks). Equivalent initiatives will also be required to build resilience against EMP threats to blackstart-essential communications. Chapter VIII examines specific options to strengthen such resilience.

Adversaries may attack the grid and other infrastructure not only to disrupt NCFs and other vital public safety and defense functions, but also to incite public panic and corrode citizen support for allied defense operations.

2. Chaos management:

Adversaries may attack the grid and other infrastructure not only to disrupt NCFs and other vital public safety and defense functions, but also to incite

78 NERC, GridEx VI Lessons Learned Report, April 2022, vi

public panic and corrode citizen support for allied defense operations. For TOPs and their personnel conducting blackstart restoration, having to operate in the face of societal chaos could greatly complicate their already challenging tasks.

Senior U.S. officials are especially concerned that such panic-inducing attacks could occur in the contest of a Chinese invasion of Taiwan. Cybersecurity and Infrastructure Security Agency (CISA) Director Jen Easterly warns that "A scenario of attacks against our critical infrastructure in the event of a Chinese invasion of Taiwan is unfortunately not far-fetched." Such an invasion "might very well be coupled with the explosion of multiple gas pipelines, the mass pollution of our water systems, the hijacking of our telecommunication systems, the crippling of our transportation nodes," she added. "All designed to incite chaos and panic across our country and deter our ability to marshal military might and citizen will."[79]

China's People's Liberation Army (PLA) will also conduct information operations to magnify the psychological impacts of blackouts and other infrastructure disruptions. The U.S. Department of Defense assesses that "prior to the start of hostilities, the PLA would conduct psychological warfare to erode the will to fight of both adversary military forces and the civilian population." As the conflict progresses, "the PLA would probably continue to use information operations against an adversary's military and civilian populations in order to continue to erode adversary support for the conflict."[80]

The blackstart restoration process itself will create public concerns and political pressures on TOPs and their partners in industry and government. Society as a whole has little awareness of the blackstart restoration process and the requirements for sequential, prioritized load pickup that may not coincide with the immediate preferences of mayors, governors, and the citizens they represent, including restoring power to residential areas. TOPs will not initially restore service to entire cities or other large blocks; the system will

79 Jeff Seldin, "Us Warns Of Massive Chinese Cyberattacks In Taiwan Scenario," Voice Of America, February 27, 2023, Https://Www.Voanews.Com/A/Us-Warns-Of-Massive-Chinese-Cyberattacks-In-Taiwan-Scenario-/6981396.Html

80 U.S. Department of Defense (DOD), Military and Security Developments Involving the People's Republic of China, November 2022, 153-4, https://media.defense.gov/2022/Nov/29/2003122279/-1/-1/1/2022-MILITARY-AND-SECURITY-DEVELOPMENTS-INVOLVING-THE-PEOPLES-REPUBLIC-OF-CHINA.PDF

need to conduct such load pickups in much smaller increments. Moreover, in an extended conflict with China or other adversaries, the Secretary of Energy may issue grid security emergency orders (as authorized by the Federal Power Act) requiring TOPs to restore service to critical defense facilities. Government leaders may also direct the prioritization of power delivery to seaports and other facilities necessary for deploying forces to regional conflict zones.

The Electricity Subsector Coordinating Council (ESCC) and its members have built an effective, frequently used system to coordinate public communications when blackouts or other major grid disturbances occur. The ESCC notes that it "maintains a strong focus on industry and government working in parallel to address national security threats, ensuring unity of effort and unity of message across the sector. The severity and nature of threats to the grid can be unpredictable. To combat this, the ESCC serves as a center of gravity to enable coordination of effective responses and aligned messaging. The Council has also developed a "playbook" that provides senior industry and government executives with a framework to coordinate response and recovery efforts and communication with the public during major incidents. The playbook is updated annually, tested in a series of exercises (most recently GridEx VI), and used during real-world events.[81]

Electric utilities and their industry and government partners should build on these playbooks to explicitly account for chaotic conditions in which blackstart restoration will likely need to go forward, especially in adversary-induced outages that are paired with information operations designed to magnify public panic and disorder. TOP training and exercises should also account for the potential challenges of operations in such an environment.

3. Additional factors to be addressed later in this Handbook for inclusion in training, simulations, and exercises:

- Disruption of natural gas flows to gas-fired generators essential for restoration;

81 Electricity Subsector Coordinating Council (ESCC), Protecting The Energy Grid From National-Level Disasters And Threats Is A Responsibility The Government And The Electric Power Industry Share, 2023, Https://Www.Electricitysubsector.Org/-/Media/Files/ESCC/Documents/ESCC_Brochure

- Disruption of other interdependent infrastructure, including transportation systems necessary for the resupply of secondary fuels;
- Shortfalls in substation battery power if multiple re-starts of the blackstart restoration process are necessary; and
- Sustaining and providing security for the personnel essential for conducting blackstart restoration, including those forward deployed to remote substations, especially if their families and communities lose water service and other lifeline functions and are subject to adversary information operations. The Transmission Resilience Maturity Model (TRMM) emphasized the need to establish and be able to execute family care plans.[82]

B. Operating in Blackout Conditions

NERC standards do not require TOPs to test the re-energizing of their cranking paths. The lack of such a requirement is understandable: to conduct realistic exercises, most utilities would need to create a power outage over a portion of their service area, and halt service to homes, businesses, health care facilities, or other customers who rely on grid-provided power. No power company wants to deprive its customers of electricity and create potential reliability issues to carry out such an exercise, especially since (in the absence of prior tests) unanticipated problems might occur and delay restoration of service. Of course, that same risk of undiscovered shortfalls in blackstart

Risk of undiscovered shortfalls in blackstart plans and capabilities makes it all the more vital to strengthen modeling and simulation assessments of restoration effectiveness.

82 North American Transmission Forum, Transmission Resilience Maturity Model, https://trmm.labworks.org/

plans and capabilities makes it all the more vital to strengthen modeling and simulation assessments of restoration effectiveness, and – to the limited extent possible – introduce greater reality into blackstart exercises and testing.

Fully realistic exercises would be problematic for many utilities due to other reasons as well. For example, they would need to place simulated malware on their networks (aka Test Effect Payloads) that might be difficult to eradicate. That malware could disrupt automation and remote management of the grid, and – in a real-world event – could reveal shortfalls in skilled operators and linemen required to man cranking path generators and substations. The COVID pandemic has introduced threats to the availability of such personnel. With a pandemic driven requirement for IT, OT, and cybersecurity staff to operate remotely, new dependencies (and perhaps attack surfaces) have emerged for remote access, maintenance, and operations of SCADA and communications networks.

> Society as a whole has little awareness of the blackstart restoration process and the requirements for sequential, prioritized load pickup that may not coincide with the immediate preferences of mayors, governors, and the citizens they represent, including restoring power to residential areas.

In the absence of realistic testing, utilities conduct simulations of blackstart restoration operations to meet EOP-005-3 requirements. The FERC-NERC-Joint Entity study found that the simulations used by TOPs and other entities participating in the study "verify the capability of blackstart facilities and of other generation resources to meet real and reactive power requirements of cranking paths and verify their dynamic capability to supply priority loads."[83] Participants also effectively used an operator (OTS) or dispatcher training simulator (DTS) in exercises to simulate blackstart conditions and operations. The OTS/DTS allows participants to exercise system restoration using the same tools and computer interfaces they encounter in

83 Federal Energy Regulatory Commission (FERC) and NERC, Report on the FERC-NERC-Regional Entity Joint Review of Restoration and Recovery Plans, January 2016, p. 46, https://www.ferc.gov/legal/staff-reports/2016/01-29-16-FERC-NERC-Report.pdf. For more details on the types of dynamic and steady-state analyses that study participants performed, see pp. 45-47.

normal operations and provides situational awareness of simulated system conditions.[84]

The study also noted, however, that provisions for blackstart testing "do not necessarily require verification using conditions that anticipate actual blackout conditions." In particular, "the blackstart testing requirements in the Reliability Standards are limited to ensuring that a blackstart unit is functional, but they do not explicitly require verification of the ability to energize equipment under the conditions anticipated during an actual blackout situation."[85] Moreover, no requirements exist for exercises to replicate a situation where adversaries have disrupted grid infrastructure and are continuing to attack it in order to obstruct blackstart restoration.

C. The DARPA RADICS Program as a Best Practice

Transmission operators and their partners should explore politically and operationally realistic options to fill these gaps. A prime example of options to do so has been provided by the Rapid Attack Detection, Isolation and Characterization Systems (RADICS) exercise program conducted by the Defense Advanced Research Projects Agency (DARPA). The agency notes that "a substantial and prolonged disruption of electric power would have profound economic and human costs for the United States. From a defense perspective, a major power outage could hamper

> The blackstart testing requirements in the Reliability Standards are limited to ensuring that a blackstart unit is functional, but they do not explicitly require verification of the ability to energize equipment under the conditions anticipated during an actual blackout situation.

84 FERC-NERC-Joint Entity Report, Review of Restoration and Recovery Plans (2016), p. 57.

85 FERC-NERC-Joint Entity Report, Review of Restoration and Recovery Plans (2016), p. 40.

military mobilization and logistics and impair the capability to project force." Accordingly, DARPA, DOE, and electric utilities structured RADICS to help enable and accelerate black start recovery of the power grid amidst a cyber-attack on the U.S. energy sector's critical infrastructure.[86]

To avoid the risks to surrounding infrastructure and electricity customers that a realistic blackstart exercise could entail, DARPA collaborated with the Department of Homeland Security (DHS), to develop a testbed on Plum Island, New York. The island provided an isolated environment for the safe construction and use of a cranking path. Particular objectives of the exercise:

Plum Island blackstart testbed objectives

a. **Enhanced situational awareness (SA).**

Improved SA can help operators thwart an attack or blunt its effects and thereby minimize physical damage to electrical equipment that would otherwise take significant time to repair. RADICS technologies will improve situational awareness by providing accurate and timely information about grid status before, during, and after an attack.

b. **Counter sustained attacks.**

An attacker may continue network-based attacks during recovery efforts to delay power restoration. RADICS technologies will enable the creation of isolated emergency networks that would permit secure responder coordination and power systems communication.

86 Defense Advanced Research Projects Agency (DARPA), Technologies to Rapidly Restore the Electric Grid after a Cyberattack Come Online, February 23, 2021, https://www.darpa.mil/news-events/2021-02-23; DARPA, Rapid Attack Detection, Isolation and Characterization Systems (RADICS), https://www.darpa.mil/program/rapid-attack-detection-isolation-and-characterization-systems. This section and other material in the Handbook also draw on RADICS-related findings and recommendations in Paul Stockton, Blackstart for National Security: Plans, Capabilities, and Exercises for the Emerging Treat Environment, May 19, 2021, a research project funded by the Defense Advanced Research Projects Agency. The views and conclusions contained in this document are those of the author and should not be interpreted as representing the official policies, either expressed or implied, of the defense advanced research projects agency or the U.S. government.

c. **Identification of grid malware.**

Cyber-attacks on critical infrastructure can take many forms, including the corruption of configuration files and introduction of malicious code. RADICS technologies will safely and automatically map and assess the state and configuration of electrical power networks and detect and characterize power grid malware.

d. **Survivable communications.**

Communications between control centers, substations personnel, and other participants in blackstart restoration are absolutely vital to that process and may be targeted for disruption accordingly. The exercise enabled the testing of the EIS-Council's BSX emergency communications system and other highly survivable emergency options.[87]

Developing and executing equivalent exercises and testbeds will be enormously helpful in strengthening blackstart resilience against Black Sky hazards. In particular, for TOPs who will employ gas-fueled generators on their cranking paths (and for growing the power islands they establish), those companies and their partners should conduct drills based on the partial or complete disruption of the natural gas pipelines that fuel such generators for multiple days. TOPs should also conduct simulations, drills, and/or TTXs to assess the blackstart implications of the expansion of IBRs, which are creating a new set of challenges for blackstart plans and operations.

87 For a comprehensive discussion of communications needs for utilities and other sectors in Black Sky Events, see EIS Council, Black Sky Communications Risks, https://eiscouncil.org/black-sky-event-communications-risk/

D. Modeling and Simulation

TOPs and their partners use increasingly advanced modeling and simulation tools to help optimize their blackstart plans and, as needed, develop and implement fallback strategies if damage to infrastructure precludes the employment of their primary plans. The use of Real Time Digital Simulations (RTDS) of the utility's electric system provides an excellent way to observe the dynamic behavior of the system during a blackstart in conjunction with the Dispatch Training Simulator (DTS). Once the system is developed for RTDS, it becomes a tool to assist the TOP in meeting real time system operation challenges.

The value of these modeling and simulation (M&S) tools for restoration support will grow as inverter-based resources are deployed more widely, and – potentially – begin to contribute to the restoration process. IBRs differ from gas-fired generators and other synchronous assets across a wide array of performance features. M&S tools will be essential to help understand the impact of these differences in the restoration process and facilitate the use of IBRs for restoration. However, these same differences will require aggressive efforts to understand and effectively model the real-world behavior of IBRs in a restoration environment. Moreover, ensuring that these modeling and simulation tools have the auxiliary power and survivable communications to function in a severely disrupted environment will be essential for Black Sky events.

E. Providing Security for Restoration Personnel

Given the goal of China and other potential adversaries to create public panic and disorder by attacking the grid, industry and government should develop and exercise plans for security support that are scaled to potential Black Sky restoration requirements. Police and State National Guard (NG) personnel routinely provide public safety and security support to utilites during restoration operations that follow severe hurricanes and other natural disasters. TOPs should partner with government officials and public safety organizations to plan and exercise for larger-scale, extended support during Black Sky outages. However, such outages will also create competing demands on law enforcement, security contractors, and NG assets to secure other facilities and maintain public order. Government leaders and industry should develop criteria in advance of Black Sky events to help allocate and sequence the delivery of scarce security resources, rather than make up such guidelines in the mist of the catastrophe.

STRUCTURE OF THE HANDBOOK

Chapter II examines the vital role that blackstart will play in Black Sky events to support societal continuity and national security, and the implications of that importance for prioritizing load pickup and overall blackstart system design. This Chapter also examines how Artificial Intelligence (AI) tools might help restoration personnel manage the complexity of such decision-making, and how advanced microgrids might reduce the number of facilities that require prioritized power restoration.

Chapter III analyzes the problems of ensuring resource adequacy for blackstart restoration in the context of 1) federal and state zero-carbon initiatives that will put increasing pressure on the availability of gas-fired generators; 2) the broader requirements for blackstart power production as IBR penetration increases; and 3) options to help meet blackstart resource needs even as the broader decarbonization of the generation fleet goes forward.

Chapter IV explores the challenges of maintaining the operating reliability of the power islands that the blackstart restoration process will create, from the energizing of the cranking path through island growth, synchronization with neighboring islands, and – ultimately – reestablishment of interconnections. This chapter also offers recommendations to provide for such operating reliability in the face of increasing IBR penetration.

Chapter V does a “deep dive” into opportunities to incorporate IBRs into the blackstart restoration process, to 1) serve as the blackstart resources that begin energizing the cranking path; 2) supplement synchronous generators for next start; or 3) to entirely replace those synchronous generators, which will only be possible if immense (and in some cases poorly understood) technical and operational problems are resolved. The chapter assesses both grid-following and grid-forming IBRs, and supplements existing roadmaps and strategies to overcome the myriad impediments to their use for blackstart.

Chapter VI accounts for the fact that today and for many years to come, natural gas-fired generators will not only be essentail for BES reliability and blackstart, but will also be at increasing risk of incurring disruptions in their supply of fuel. The chapter highlights specific threats to natural gas infrastructure and proposes an array of collaborative measures by BES entities and gas system operators to strengthen the resilient flow of fuel in Black Sky events.

Chapter VII offers an overview of additional zero-carbon sources of power that might support blackstart restoration, including: small-scale hydropower, geothermal, fuel cells, and small nuclear reactors. In addition, the chapter explores options for distribution grid blackstart (DGBS), and highlights the benefits of diversifying the generation fleet to complicate adversary attack planning and execution.

Chapter VIII proposes a range of additional measures including asset hardening, sparing, emergency communications, and chaos management to strengthen blackstart infrastructure resilience against all Black Sky threats.. Finally, the chapter puts these efforts into the larger, important context of human continuity.

2

CHAPTER TWO

THE IMPORTANCE OF BLACKSTART FOR SOCIETAL CONTINUITY AND NATIONAL SECURITY

INTRODUCTION

The imperative for resilient blackstart restoration plans and capabilities reflects the complete dependence of modern life and the economy on electric power, and the vital role that electricity plays in national security for the United States and its international partners. The ongoing growth of these dependencies, and adversary efforts to exploit them, provide important design parameters for shaping future blackstart resilience against Black Sky hazards.

Section B of this chapter examines how electrification of the economy and increasingly complex interdependencies between infrastructure sectors will create new challenges for power restoration. This section also proposes options to help account for these factors, including 1) electric industry participation in National Risk Management Center's development of a National Critical Function Risk Architecture, and 2) the establishment of new consultative mechanisms to bring restoration stakeholders together to help ensure societal continuity in Black Sky events.

Section C explores how Artificial Intelligence (AI) tools can help map these sector interdependencies and refine existing restoration plans. This section also examines whether and how generative AI might provide decision support to grid operators conducting blackstart in a severely disrupted environment – including in the face of the sophisticated attack planning and execution that AI will make possible for our adversaries.

Section D analyzes the prospects and limitations for employing emergency power systems to sustain the functionality of especially critical facilities, and thereby reduce the number of such facilities that will require prioritized blackstart restoration. In addition, this section examines options to expand the deployment of large-scale microgrids that employ long duration energy storage systems, and the significant technical challenges that must be resolved before their wide-scale deployment becomes possible.

Section E explores the growing importance of blackstart for national security. DOD and DHS officials warn that in a conflict over Taiwan, China may attack the U.S. grid to disrupt the flow of forces from stateside military installations to the ports from which they will embark to the conflict zone. China may also seek to disrupt other sectors to create Black Sky-scale threats to U.S. public health and safety, and thereby corrode public support for coalition defense and intensify pressure on the U.S. President to yield in the conflict. In scenarios of this sort, effective blackstart capabilities will be crucial for national security. Section E examines the implications of such homeland defense challenges for blackstart system design and operations.

Section F analyzes a specific option to meet these challenges: how the U.S. and its international partners can utilize their governments' emergency authorities to help guide blackstart restoration. In particular, the section examines how the Secretary of Energy could issue Grid Security Emergency Orders to help prioritize power restoration operations for homeland defense. This section also examines the failures that are likely to occur unless joint industry-government planning and exercises build preparedness for emergency order execution.

I | BLACKSTART TO SUPPORT SOCIETAL CONTINUITY

Previous EPRO Handbooks have documented the massive dangers that Black Sky outages would pose to public health and safety. For example, if extended, multi-state outages inflicted multi-week disruptions of electricity-dependent water and wastewater utilities, the consequences for public health and safety would be catastrophic.[88]

TOPs have detailed plans to prioritize the restoration of service to these systems. However, the water sector is only one of many that are vital for societal continuity. Moreover, these sectors are increasingly interdependent, with data centers and other new (and very large) types of loads competing for prioritized restoration against many other critical assets and functions. Restoration plans

88 EIS Council, E-Pro Handbook II, Volume II: Water Sector Resilience for Black Sky Events

will need to account for the growing diversity of such customers and the difficulty of averting cascading failures between them.

A. Community Lifelines

The water sector is only one of many community lifeline infrastructures that will be in jeopardy in Black Sky events. The Federal Emergency Management Agency (FEMA) defines these lifelines as "the most fundamental services in the community that, when stabilized, enable all other aspects of society to function." Each such lifeline also enables "the continuous operation of critical government and business functions and is essential to human health and safety or economic security."[89] Every single one of these community lifelines depends on the availability of electric power. Failure to embed prioritized, robust restoration capabilities for such lifelines in blackstart planning would leave society vulnerable to mass casualty events.

The water sector is only one of many community lifeline infrastructures that will be in jeopardy in Black Sky events. The Federal Emergency Management Agency (FEMA) defines these lifelines as "the most fundamental services in the community that, when stabilized, enable all other aspects of society to function.

In the United States, FEMA identifies the following as critical lifeline infrastructures:

- **Safety and Security** - Law Enforcement/Security, Fire Service, Search and Rescue, Government Service, Community Safety
- **Food, Hydration, Shelter** - Food, Hydration, Shelter, Agriculture
- **Health and Medical** - Medical Care, Public Health, Patient Movement, Medical Supply Chain, Fatality Management

89 Federal Emergency Management Agency (FEMA), Community Lifelines, https://www.fema.gov/emergency-managers/practitioners/lifelines

- **Energy** - Power Grid, Fuel
- **Communications** - Infrastructure, Responder Communications, Alerts Warnings and Messages, Finance, 911 and Dispatch
- **Transportation** - Highway/Roadway/Motor Vehicle, Mass Transit, Railway, Aviation, Maritime
- **Hazardous Materials** - Facilities, HAZMAT, Pollutants, Contaminants
- **Water Systems** - Potable Water Infrastructure, Wastewater Management

Beyond FEMA's list, a growing number of other infrastructure assets and operations have become central to modern economies, all of which are dependent on electricity and constitute emerging priorities for power restoration planning. Two infrastructure sectors help illuminate the broader challenges for TOPs and their partners:

- **Financial Services Sector**

This sector is included under communications in FEMA's list of community lifelines. But that listing falls short of reflecting the immense scale, complexity, and societal importance of financial services. CISA emphasizes that "the Financial Services Sector represents a vital component of our nation's critical infrastructure," and that "large-scale power outages" constitute one of the key potential risks facing the sector. CISA also notes that the sector "includes thousands of depository institutions, providers of investment products, insurance companies, other credit and financing organizations, and the providers of the critical financial utilities and services that support these functions."[90]

Given the rise of "fintech" – defined as "technology that seeks to improve and automate the delivery and use of financial services."[91] – and the rapid evolution of the sector as a whole, determining which of these assets should receive prioritized power restoration (as opposed to being part of large load blocks to be picked up in the latter phases of restoration) is an increasingly complex and high stakes undertaking.

90 CISA, Financial Services Sector, https://www.cisa.gov/topics/critical-infrastructure-security-and-resilience/critical-infrastructure-sectors/financial-services-sector

91 "What is Financial Technology (Fintech)? https://www.investopedia.com/terms/f/fintech.asp

- **Data centers and the broader Information Technology (IT) Sector.**

CISA states that the Information Technology Sector "is central to the nation's security, economy, and public health and safety as businesses, governments, academia, and private citizens are increasingly dependent upon IT Sector functions. These virtual and distributed functions produce and provide hardware, software, and IT systems and services, and—in collaboration with the Communications Sector—the Internet."[92]

It is impossible to overstate the importance of the internet and IT services to the evolving power grid. Distributed energy resources (DERs), energy aggregators, and other growing components of U.S. and other nation's electric systems are heavily dependent on internet-based communications. Many other grid functions rely on IT services as well. In a Black Sky outage, all these support functions would be disrupted. TOPs typically rely on their own data processing and simulation capabilities to support power restoration. However, as interdependencies between the IT sector and the electric system continue to evolve, and as IT-related loads become increasingly important to society writ large, new challenges will emerge for blackstart operations.

Data centers provide a case in point. U.S. data power consumption is expected to reach 35 gigawatts by 2030, up from 17 gigawatts in 2022. While data centers vary in size, they represent very large loads, typically requiring between 100-300 megawatts to function.[93] Virtually all these centers have robust emergency power generators. However, as will be discussed below, sustaining the resupply of fuel to these generators in a long duration blackout will become increasingly difficult and ultimately impossible, given the competing demands for fuel and the disruption of transportation systems that such an outage would create.

Restoring grid-provided power to data centers will be imperative, given the role they now play in the economy and the functioning of so many critical infrastructure sectors. But the massive loads that each

92 CISA, Information Technology Sector, https://www.cisa.gov/topics/critical-infrastructure-security-and-resilience/critical-infrastructure-sectors/information-technology-sector

93 Todd Forrest, "Rethinking Data Center Power," May 8, 2023, Microsoft Data Interconnection Center,https://www.hdrinc.com/insights/rethinking-data-center-power

of these centers represent, and their aggregate loads in areas such as Northern Virginia, will pose significant challenges for sequencing restoration operations in the face of multiple competing requirements to power community lifelines and other critical facilities and functions.

B. Electrification: Implications for Blackstart and the Prioritized Restoration of Power

As part of the broader drive for carbon reductions, the United States and many other nations are transitioning away from the use of fossil fuels in multiple sectors, and shifting instead to electricity (generated, ideally, by zero carbon sources of energy). The electrification of manufacturing and other industrial processes will create major new demands for electric power.[94] The growth of electric vehicles at the expense of their gas-fueled counterparts is accelerating at an extraordinary rate as well, with some projections estimating that U.S. sales of EVs will surpass 50% of all vehicle sales by 2050.[95] The International Energy Agency estimates that equally rapid growth will occur in many nations around the globe.[96] These trends have profound implications for future restoration requirements and risks.

> As more and more critical functions and assets depend on electricity rather than gas or other fossil energy sources.

94 DOE, Industrial Decarbonization Roadmap, September 2022, https://www.energy.gov/sites/default/files/2022-09/Industrial%20Decarbonization%20Roadmap.pdf; Optimas, The Future of Industrial Processes Is Electrification, https://optimas.com/blog/the-future-of-industrial-processes-is-electrification/

95 U.S. Bureau of Labor Statistics, Charging into the future: the transition to electric vehicles, February 2023, https://www.bls.gov/opub/btn/volume-12/charging-into-the-future-the-transition-to-electric-vehicles.htm#:~:text=S%26P%20Global%20Mobility%20forecasts%20electric,surpassing%2050%20percent%20by%202030.

96 International Energy Agency, Global EV Outlook 2023 – Analysis, https://www.iea.org/reports/global-ev-outlook-2023/executive-summary

As more and more critical functions and assets depend on electricity rather than gas or other fossil energy sources, two restoration-related issues will emerge. First, the total load to be restored will greatly expand, and accelerating the restoration process will be even more important than it is today to save and sustain lives and maintain economic and societal continuity. Prioritizing load restoration between critical customers will become more challenging as well. Second, as electrification goes forward, the speed and scale of cascading failures across electricity-dependent facilities and functions will increase.

Building electrification helps illustrate these trends and their implications for restoration. Initial construction and the ongoing use of buildings and their HVAC and related systems, both commercial and residential, produce about 40 percent of carbon emissions in the U.S.. These emissions stem from the mining, processing, manufacture, transportation, and installation of the materials buildings are made of, and from the operation of their heat, electric, and cooling systems, cooking, and appliances.[97]

A growing number of U.S. cities and states are implementing new laws to drive the electrification of new construction and eventually all buildings. In 2021, for example, California adopted a building code that incentivizes builders to go all-electric for heating and appliances. Berkeley, Calif. had already banned natural gas in new buildings in 2019, followed by San Francisco. Seattle has prohibited most uses of fossil fuels in new buildings in 2021. Denver, Colo. is requiring large buildings to shift to electric heating by 2025. Ithaca, N.Y. is aiming to decarbonize all its buildings by 2030. In New Jersey, all buildings must be fully electric by 2050. And in New York state and New York City, especially stringent requirements for electrification have been enacted for heating, cooling, and other functions across a broad range of buildings.[98]

Transportation systems, and many other sectors and economic activities are undergoing rapid electrification as well. The net effect: major increases are likely in the total demand for (and dependence on) electric power. The National Renewable Energy Laboratory (NREL) estimates that widespread electrification

97 Columbia Climate School, What Makes Electrifying the Economy So Challenging, June 3, 2022, https://news.climate.columbia.edu/2022/06/03/what-makes-electrifying-the-economy-so-challenging/

98 Ibid

could boost U.S. electricity consumption by almost 40 percent by 2050.[99] A variety of factors could alter this demand trajectory, including the rate of overall growth of the U.S. and allied economies. Nevertheless, from a grid planning perspective, the prospect of such significant growth in power consumption is already driving reassessments of resource adequacy and new transmission requirements (including those associated with the growth of solar and wind resources that are remote from major loads).

The implications for blackstart planning are at least as consequential. As society and the economy become increasingly reliant on electric power, the potential consequences of Black Sky power outages will grow as well. Again, building heating and cooling systems exemplify these consequences. Such systems would not typically be considered priority loads for blackstart restoration. Indeed, given the sequenced, incremental process of load pickup built into most restoration plans, the very large loads represented by urban areas and their residential building would typically be restored to service relatively late in the restoration process. But as climate change continues and weather events (both cold and hot) become more extreme and frequent, the health and safety consequences of long duration, large scale outages in terms of building heating and cooling could be catastrophic.

Future blackstart plans and capabilities will need to account for these interdependencies in ways that go above and beyond traditional criteria for prioritized restoration.

As noted in the introduction, TOPs already have detailed plans to prioritize and sequence the restoration of power to customers that are vital for societal continuity, including water systems, nuclear power plants and other community lifelines. Electrification will create much broader demands for accelerated power restoration. So, too, will the risk of cascading failures between infrastructure sectors, where facilities that may not be critical within their own sectors are necessary for sustaining others. Future blackstart plans and capabilities will

99 National Renewable Energy Laboratory (NREL), NREL Analysis Explores Demand-Side Impacts of a Highly Electrified Future, July 9, 2018, https://www.nrel.gov/news/program/2018/analysis-demand-side-electrification-futures.html

need to account for these interdependencies in ways that go above and beyond traditional criteria for prioritized restoration.

1. National Critical Functions

These same recommendations are also applicable for the facilities and interdependent services essential to sustaining the continuity of National Critical Functions (NCFs) in Black Sky events. As noted earlier in this Handbook, NCFs constitute "the functions of government and the private sector that are so vital to the United States that their disruption, corruption, or dysfunction would have a debilitating effect on security, national economic security, national public health or safety."[100]

Plans to prioritize the restoration of service to NCFs must account for the incredibly complicated nature of their sub-functions and dependencies. CISA's National Risk Management Center (NRMC) development of an NCF Risk Architecture will help build out that understanding. That architecture is designed to identify the key assets, networks and systems that support those functions, as well as the underlying software, hardware, and other technologies that enable that process. The NRMC's initiative will also identify the entities – whether businesses or government – that are key providers of elements of the process and function. This architecture relies on process and engineering maps, as well as understanding of business and governance of the function to enable risk analysis against a range of scenarios.[101]

TOPs and their partners should begin discussions with CISA and the NRMC on leveraging the NCF Risk Architecture to support blackstart restoration planning. Indeed, absent a government-provided assessment of sub-function loads and their complex interdependencies, restoration planners would face the incredibly burdensome task of doing so themselves. The ESCC, NATF, and other industry organizations should reach out to the leadership of the NRMC to initiate such discussions.

100 CISA, National Critical Functions, https://www.cisa.gov/topics/risk-management/national-critical-functions#:~:text=National%20Critical%20Functions%20(NCFs)%20are,safety%2C%20or%20any%20combination%20thereof.

101 National Critical Functions, 2

Yet, existing restoration plans and the expertise of the electricity subsector will also be vital to the success of NRMC's effort. The success of the Architecture will depend on the willingness of asset owners and operators to provide information that they may consider business-sensitive or proprietary. The NRMC states that it "will continue developing this information in coordination with critical infrastructure stakeholders, endeavoring to deepen the understanding of who and what is required for the successful, sustained, and resilient operation of individual NCFs." The Center also notes that "this is a complex and ongoing undertaking and will rely heavily on the expertise of the critical infrastructure community."[102]

The operators of the power grid, as well as the operators of natural gas systems and other interdependent infrastructure essential for power restoration, can and should make immense contributions to this effort. Especially valuable: TOPs and their industry partners have a unique understanding of the layout of their generation, transmission, and distribution systems, and of the sequencing of the restoration process, including restoration of critical loads necessary for that process to go forward. Planning for restoration of priority load cannot occur in an engineering (much less a physics) vacuum; industry input will be vital to align those priorities with the constraints of the restoration process.

Planning for restoration of priority load cannot occur in an engineering (much less a physics) vacuum; industry input will be vital to align those priorities with the constraints of the restoration process.

Blackstart stakeholders (including the electricity subsector, emergency managers, regulators, and elected officials) should also adopt new consultative mechanisms and organizations to help TOPs shape and refine their restoration plans. Many TOPs, RCs, and other participants in the restoration process already discuss blackstart planning with State, local, tribal, and territorial (SLTT) emergency managers and other incident response officials. However, given the diversity and complex supporting sub-functions of NCFs that will require

102 Ibid.

restoration, it would be useful to broaden and regularize these consultative mechanisms.

One option would be to establish cross-sector blackstart restoration coordinating councils to help TOPs account for 1) the implications of electrification on the diversity and overall scale of critical loads, and 2) the expanding interdependences and risks of cascading failures across community lifelines, NCFs and the critical infrastructure sectors. Such councils could align to include the key restoration stakeholders within a TOP's service area. Alternatively, to keep the size of these councils manageable, they could be responsible for smaller portions of those service areas. Under either approach, it will also be useful to support such coordination efforts with powerful new analytic and decision support tools, especially those enabled by generative AI.

However, the input that these councils provide should remain advisory. The North American Electric Reliability Corporation (NERC) assigns specific responsibilities to TOPs, RCs, and other subsector entities for blackstart planning, training, and execution, including the responsibilities and functions specified in EOP-005-3 – System Restoration from Blackstart Resources.[103] Accordingly, while coordinating councils can provide enormously valuable input on restoration priorities, TOPs and their NERC-designated partners must continue to have the final say over blackstart planning and execution.

103 North American Electric Reliability Corporation (NERC), EOP-005-3 – System Restoration from Blackstart Resources, effective April 1, 2019, R6, https://www.nerc.com/pa/Stand/Reliability%20Standards/EOP-005-3.pdf.

II | ARTIFICIAL INTELLIGENCE AS ENABLER OF BLACKSTART RESTORATION – AND AN INTENSIFYING THREAT

Grid operators already employ real time digital simulation (RTDS) and other decision support tools to help them assess alternative blackstart plans, understand and adjust to infrastructure disruptions, and optimize restoration operations. For example, Dominion Virginia Power employs an RTDS-based methodology to study the black start process on specific cranking paths in Dominion's Service Restoration Plan (SRP) in ways that account for system dynamics and relay configuration.[104] Energy Management Systems (EMS)

104 Runi Sun, Blackstart Study Using RTDS in Dominion Virginia Power, June 9, 2020, https://knowledge.rtds.com/hc/en-us/article_attachments/360074874273

also help grid operators model their network operations (including topology processing and state estimation), conduct security assessments, perform automatic generation control and dispatch, and – increasingly – optimize the use of battery energy storage systems.[105]

AI tools are now providing additional opportunities for modeling and simulation (M&S) capabilities that could be very valuable for supporting restoration operations. DOE and its national laboratories are exploring AI-enabled optimization models that simulate the electric system and the severity of disruptive events much more quickly than is possible with current widely used models.[106] Vendors are also selling AI products that provide real-time monitoring of electric systems, conduct predictive intelligence to identify system failures, and diagnostics on the defective line to help operators remedy faults and avoid cascading failures.[107] All such functions could be useful for blackstart. DOE, utilities, and product providers should make restoration a key focus of AI research and development efforts.

AI could also be helpful for refining TOP plans for prioritized restoration. Given the growing number and complex interdependencies of NCF and community lifeline loads, and the immense amount of (frequently changing) data required to support such restoration planning, it will become increasingly difficult for humans to optimize restoration for societal continuity without machine support. The National Security Commission on Artificial Intelligence found that AI-enabled tools have the potential to augment filtering, flagging, and triage across multiple

DOE, utilities, and product providers should make restoration a key focus of AI research and development efforts.

105 Emerson, Battery Energy Management Systems, https://www.emerson.com/en-us/automation/control-and-safety-systems/distributed-control-systems-dcs/ovation-distributed-control-system/battery-energy-management-system?gclid=Cj0KCQjw1OmoBhDXARIsAAAYGSHMBwfoWBfzYF9vEU1R6yP5BSC23hpMnfsmnbdpESBs-FK780_hCIQaAkg8EALw_wcB

106 DOE, Artificial Intelligence Can Make the U.S. Electric Grid Smarter and More Reliable, https://www.energy.gov/technologytransitions/articles/artificial-intelligence-can-make-us-electric-grid-smarter-and-more

107 V-Soft, Secure and Maintain Power Grids Effectively With Artificial Intelligence, https://blog.vsoftconsulting.com/blog/secure-and-maintain-power-grids-with-artificial-intelligence

data sets." Such tools can also "identify connections and correlations more efficiently and at a greater scale than human analysts and can flag those findings and the most important content for human analysis."[108] All such capabilities can support blackstart planning in ways that optimize the pickup of complex, interdependent loads for societal continuity.

The Department of Defense is pioneering the development of AI systems that can manage and exploit large data sets for planning, mapping and modeling, and real-time decision support. The National Geospatial-Intelligence Agency (NGA) provides a case in point. Every second, some 70,000 new data points flow into the Agency's cybersecurity center. "The amount of data that we are bringing in from a cyber perspective is absolutely astounding," said Gary Buchanan, the NGA's chief information officer. "And there isn't a workforce large enough to sift through that level of data every day, which is why we really need artificial intelligence and machine learning for our future." Especially significant, Defense agencies are developing generative AI tools that will provide text, images, and even audio input for decisionmakers in response to user prompts.[109] With the help of DOD, utilities and their partners should explore whether and how equivalent generative tools may be able to support blackstart planning and – over the longer term – real-time decision support for restoration operations.

2. Managing AI Challenges

However, deep problems surround the veracity of AI-generated analysis and recommendations. Pentagon leaders warn that because AI large language models sometimes produce false or unsupported findings (i.e., "hallucinate"), careful decisions must be made about the uses that are appropriate for those models.[110] A U.S. Government Accountability Office (GAO) report on AI also

108 National Security Commission on Artificial Intelligence, Final Report, February 2021, https://www.nscai.gov/wp-content/uploads/2021/03/Full-Report-Digital-1.pdf

109 Lauren Williams, "Swamped with cybersecurity data, NGA hopes ChatGPT-like tools can help," Defense One, September 29, 2023, https://www.defenseone.com/defense-systems/2023/09/swamped-cybersecurity-data-nga-hopes-chatgpt-tools-can-help/390808/

110 Patrick Tucker, The Pentagon just launched a generative AI task force, Defense One, August 10, 2023, https://www.defenseone.com/technology/2023/08/defense-department-just-launched-generative-ai-task-force/389298/

cautions that "automated systems themselves are susceptible to a range of disruptive and deceptive tactics that might be difficult to anticipate or quickly identify. These threats are amplified by the ongoing delegation of decision-making, sensing, and authentication roles to potentially vulnerable automated systems. Moreover, broader deployment could become riskier as the reliance on autonomous decision-making increases."[111] Developing human operator trust of AI-generated recommendations poses an additional challenge, whether in fighter cockpits or in grid control centers. TOPs and their partners must take all such factors into account in using AI tools to refine and, potentially, help execute blackstart operations.

DOE and the electric industry must also more fully account for the ways in which adversaries can exploit AI to plan and execute cyberattacks on the grid, both to create Black Sky outages and to disrupt restoration efforts. The National Security Commission on AI found that "the expanding application of existing AI cyber capabilities will make cyberattacks more precise and tailored, further accelerate and automate cyber warfare, enable stealthier and more persistent cyberweapons, and make cyber campaigns more effective on a larger scale."[112]

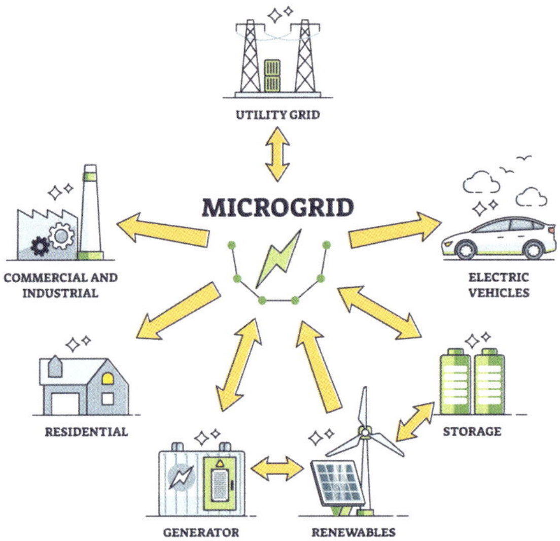

111 Government Accountability Office (GAO), Information Environment Opportunities and Threats to DOD's National Security Mission, September 2022, https://www.gao.gov/assets/gao-22-104714.pdf

112 National Security Commission on AI, 51

It would be prudent to assume that China and other highly AI-capable adversaries are gathering all available data (including from open source and satellite imagery) to carefully map all primary and backup cranking paths in the U.S. and partner nations, and use that data to assemble detailed, utility-specific attack plans. We should also assume that if a conflict occurs over Taiwan or some other regional flashpoint, China will use AI to conduct such attacks at machine speed on a nationwide scale.

Defense against such AI-enabled cyberattacks is likely to make AI-enabled grid defense all the more necessary, warts and all. The Commission emphasized that "defending against AI-capable adversaries operating at machine speeds without employing AI is an invitation to disaster. Human operators will not be able to keep up with or defend against AI-enabled cyber or disinformation attacks, drone swarms, or missile attacks without the assistance of AI-enabled machines."[113] In some ways, the adoption of AI-supported tools for grid defense is an extension of existing uses of EMS and other tools to manage grid operations. But in other ways, including the execution of blackstart plans in the face of AI-designed and executed cyberattacks on restoration infrastructure, we are just at the beginning of a vital but difficult journey.

---

113 National Security Commission on AI, 9

III | EMERGENCY POWER AND ADVANCED MICROGRIDS FOR NCFs AND COMMUNITY LIFELINES

In blackouts lasting up to a few days, diesel-fueled emergency power generators have often (though not always) been able to sustain operations by hospitals, water systems and other community lifeline facilities. These emergency capabilities will be totally inadequate in Black Sky outages. A more promising opportunity to sustain essential services is now emerging: the development of microgrids built around "hybrid" power plants that combine large-scale solar or wind generation with long duration battery energy storage systems (BESS).

Over time, the deployment of these advanced microgrids at especially critical facilities could help TOPs manage competing demands for prioritized blackstart restoration. But significant challenges remain in providing affordable, effective long-duration storage. Moreover, as demonstrated by Defense microgrids that rely on fossil generation, the ability of these installations to blackstart their own electric systems after segmenting from the grid remains a significant challenge.

Microgrids have the potential to serve two additional functions, beyond the provision of backup power for critical facilities. First, grid-leading or grid-forming microgrids can form the core of a "minimum grid" used to initiate the crank chain for blackstart. Defense microgrids are ideally suited for this function because defense facilities are well protected, centrally managed, and scattered widely across the power grid. Second, grid-forming microgrids can be placed at the edges of "minimum grid" islands to function as grid-scale protection against catastrophic failures that cascade through the power transmission system. It may also be helpful to place grid-forming microgrids at critical nodes in the power transmission network to act as protection against cascades and also for blackstart cranking purposes. Network theory now provides the tools necessary to identify these critical nodes and optimize grid-forming microgrid placement.[114], [115]

> A more promising opportunity to sustain essential services is now emerging: the development of microgrids built around "hybrid" power plants that combine large-scale solar or wind generation with long duration battery energy storage systems (BESS).

A. Emergency Power Generation

Many lifeline facilities are meeting increasingly stringent standards for emergency power and offsite fuel storage to keep generators running for up to 96 hours at certain hospitals and other critical facilities.[116] However, given the

114 Mouco, A., et al. Resilience of Power Grids to Catastrophic Cascading Failures, Proc. CIGRE U.S. National Committee 2022 Grid of the Future Symposium. 2022.

115 Mouco, A., et al. A Discussion of Resilience in Power Grids, Proc. CIGRE U.S. National Committee 2023 Grid of the Future Symposium, 2023.

116 See, for example, the National Fire Protection Association Life Safety Code (NFPL) 110, Standard for

timelines required in Black Sky outages to restore grid service to these facilities, even these longer-lasting generation capabilities will fall far short of need. The amount of time required to complete the blackstart restoration process will vary with the amount of damage to restoration infrastructure in the precipitating event and in potential follow-on disruptions. That said, it could take weeks or even months to fully re-establish the interconnections and restore service to all customers.[117] Continuous fuel resupply (and replacement of generators that break down) will be required to keep emergency generators operating for such extended periods.

Many lifeline facilities are meeting increasingly stringent standards for emergency power and offsite fuel storage to keep generators running for up to 96 hours at certain hospitals and other critical facilities.

Sustaining such resupply operations would be impossible. One reason is that Black sky outages will create an overwhelming mismatch between the capacity of fuel contractors to resupply such customers and the demand for them to do so.[118] Resupply operations will also need to occur in the face of

Emergency and Standby Power Systems, https://www.nfpa.org/codes-and-standards/all-codes-and-standards/list-of-codes-and-standards/detail?code=110

117 Enhancing the Resilience of the Nation's Electricity System (2017); "Restoration following a major earthquake is a massive problem requiring a wide range of difficult engineering and construction projects in a compromised environment, with competition from other restoration priorities. For example, key bridges or roads required to access damaged facilities may be impassable. If an earthquake destroys key generating, substation, or transmission equipment, it may take weeks or months to restore service;' p. 66;

"As noted in Terrorism and the Electric Power Delivery System, a terrorist attack carried out in a carefully planned way by people who knew what they were doing could "deny large regions of the country access to bulk system power for weeks or even months. An event of this magnitude and duration could lead to turmoil, widespread public fear, and an image of helplessness that would play directly into the hands of the terrorists. If such large extended outages were to occur during times of extreme weather, they could also result in hundreds or even thousands of deaths due to heat stress or extended exposure to extreme cold" (NRC, 2012);" p. 140.

118 One notable exception: for Air Force bases and other installations that have extensive storage of jet propulsion (JP) fuel and emergency power generators capable of using such fuel, needs for external resupply of fuel by commercial or DOD-owned transportation assets is limited. David Schill, Improving Energy Security for Air Force Installations, PhD dissertation, Pardee RAND Graduate School, 2015, p. 24, https://apps.dtic.mil/dtic/tr/fulltext/u2/1000036.pdf.

blackout-induced transportation and communications outages, complicating resupply operations, as well as other obstacles posed by widespread blackouts. Emergency generator operators will be competing for resupply against many thousands of other customers in their own states and neighboring regions. Moreover, as in the case of cell towers and gas compression stations upon which the power grid inter-depends, resupply can and should be prioritized to facilities that are essential to support blackstart restoration, even as other critical customers begin to lose emergency power. Additionally, onsite fuel storage at these facilities should be sized with a power grid blackstart in mind.

The occurrence of a severe cold weather event during a Black Sky outage could further increase demand for fuel, including for household and business heating. Of course, geopolitical factors (and not climate conditions) will play the decisive role in determining whether and when adversaries will strike U.S. and allied electric systems. Nevertheless, in assessing Black Sky preparedness and the adequacy of backup generation capabilities, emergency managers and their energy sector partners should account for the "worst case" possibility that severe winter storms or intense heat waves will coincide with grid attacks and subsequent blackstart operations.

Government and industry leaders should also strengthen planning to allocate scarce fuel supplies. The National Governors Association (NGA) identifies a critical component of such planning: "identify the facilities with the highest need." The Association provides guidance for categorizing such facilities in healthcare, and implications for ensuring their energy resilience, that are broadly applicable to other lifeline sectors. In particular, the NGA recommends that to shape emergency power strategies, the planners should account for a worst case, ""black sky event' that can have cascading effects for healthcare facilities" and supporting infrastructure.[119]

Sector Risk Management Agencies (SRMAs) should also partner with the asset owners and operators in their respective sectors to develop sector-specific guidelines, as the Environmental Protection Agency has done for water and

119 National Governors Association, Prioritizing Resilience: Best Practices On Energy Resilience For Healthcare Facilities, May 11, 2023, https://www.nga.org/publications/prioritizing-resilience-best-practices-on-energy-resilience-for-healthcare-facilities/

wastewater systems.[120] Most important: to ensure that fuel distributors and other components of the resupply system are fully engaged in preparedness planning, and can help build preparedness for the immense problems that resupply operations would encounter, the Oil and Natural Gas Subsector Coordinating Council (ONG-SCC) and its members should update the 2016 Oil and Natural Gas Industry Preparedness Handbook and associated guidelines to provide recommendations on meeting these challenges.[121]

The emergency management community should also support such efforts. In 2017, DHS issued the Power Outage Incident Annex to the Response and Recovery Federal Interagency Operational Plans, which provided significant, actionable recommendations on emergency power and fuel resupply issues.[122] SLTT government officials and emergency managers should partner with fuel companies to help implement those measures in their respective jurisdictions. FEMA and the Defense Logistics Agency (DLA) can also support fuel resupply and emergency generator deployments (though many of the capabilities they bring to bear draw on the same limited set of private sector providers). And for Air Force bases and other installations that have extensive storage of jet propulsion (JP) fuel and emergency power generators capable of using such fuel, needs for external resupply of fuel by commercial or DOD-owned transportation assets is limited.[123] Nevertheless, while careful prioritization and operational planning can help sustain the critical functions at selected facilities, Black Sky outages and associated cross-sector infrastructure disruptions will eventually put all such functions at risk.

120 Environmental Protection Agency, Is Your Water or Wastewater System Prepared? What You Need to Know About Generators, March 27, 2023, https://www.epa.gov/sites/default/files/2016-03/documents/waterwastewatersystemgeneratorpreparedness.pdf

121 American Petroleum Institute, Oil and Natural Gas Industry Preparedness Handbook, April 4, 2016, http://ongsubsector.com/documents/ONG-Handbook-2016.pdf

122 DHS, Power Outage Incident Annex to the Response and Recovery Federal Interagency Operational Plans, June 2017, https://www.fema.gov/sites/default/files/2020-07/fema_incident-annex_power-outage.pdf

123 David Schill, Improving Energy Security for Air Force Installations, PhD dissertation, Pardee RAND Graduate School, 2015, p. 24, https://apps.dtic.mil/dtic/tr/fulltext/u2/1000036.pdf.

B. Advanced Microgrids

The deployment of advanced microgrids that can support single NCFs or multiple co-located facilities provides a better option for Black Sky resilience than traditional emergency power capabilities. In particular, a growing number of demonstration projects are built around hybrid plants that pair multi-megawatt solar and wind generation assets with BESS to provide extended facility power when microgrids segment from the grid and conduct their own blackstart to initiate islanded operations by their "inside the fence line" electric systems. Grid-forming or grid-leading microgrids are needed to enable islanded microgrids to interrupt cascading failures on the power grid and also to utilize microgrids in the minimum grid cranking path for blackstart outside the fence.

A growing number of demonstration projects are built around hybrid plants that pair multi-megawatt solar and wind generation assets with BESS to provide extended facility power when microgrids segment from the grid and conduct their own blackstart to initiate islanded operations.

The Department of Defense has especially far-reaching efforts underway to develop and construct such renewable-based microgrids, as well as exploring the use of advanced geothermal technologies and other options for sustained generation.[124] In addition, DOD is seeking to deploy small modular reactors and other advanced nuclear designs to ensure power resilience at remote military bases. In September 2023, for example, the U.S. Air Force awarded a contract to establish a nuclear-powered microgrid at Eielson Air Force Base southeast of Fairbanks, Alaska.[125]

124 10 USC 2911: Energy policy of the Department of Defense, https://uscode.house.gov/view.xhtml?req=granuleid:USC-prelim-title10-section2911&num=0&edition=prelim

125 Rod Walton, "U.S. Air Force Selects Oklo Microreactor for Power Project at Alaskan Base," Energy Tech, September 28, 2023, https://www.energytech.com/distributed-energy/article/21274573/us-air-force-selects-oklo-microreactor-for-power-project-at-alaskan-base?utm_source=ET+Transition&utm_medium=email&utm_campaign=CPS230928179&o_eid=4708I6058334B1I&rdx.ident[pull]=omeda|4708I6058334B1I&oly_enc_id=4708I6058334B1I

Major impediments exist to moving from demonstration projects to wide-scale adoption of advanced microgrids. Commercial deployment of long-duration energy storage systems remains an especially significant problem. For microgrids that rely on intermittent power from solar and wind arrays, such storage capabilities will be vital for sustaining critical functions in a Black Sky outage.

DOE's "Long Duration Storage Shot" establishes a target to reduce the cost of grid-scale energy storage by 90% for systems that deliver 10+ hours of duration within the decade. This initiative includes support for all types of technologies – whether electrochemical, mechanical, thermal, chemical carriers, or any combination that has the potential to meet the necessary duration and cost targets for grid flexibility.[126]

Efforts to commercialize these new technologies are also underway. DOE's Commercial Liftoff effort is structured to incentivize the development and manufacturing of long duration energy storage systems (LDES) – defined as systems that can provide 10 to 160 hours of power dispatch. DOE has concluded that to be commercially viable for large-scale deployment, technology costs must decline by 45% to 55% by 2028 to 2030 relative to costs reported now by leading LDES technologies, and the performance and working lifetime of LDES technologies must improve to compete with lithium-ion storage and hydrogen, according to the DOE report.[127]

Funding these development efforts must be a top priority for Black Sky preparedness. LDES systems can not only enable critical facilities to continue functioning, but also – as will be discussed in Chapter V – can serve as blackstart resources when paired with advanced inverters and other power electronics. Unless and until commercialization of LDES becomes practical, however, their use to support Black Sky-capable microgrids will remain limited.

An additional problem exists for microgrids that rely on diesel or other fossil-fueled generators, even when (as on military bases) unusually large fuel storage facilities are available. As with all blackstart operations, starting up the

126 DOE, Long Duration Storage Shot, https://www.energy.gov/eere/long-duration-storage-shot

127 DOE, The Pathway to Long Duration Energy Storage Commercial Liftoff, 2023, https://liftoff.energy.gov/long-duration-energy-storage/

microgrid after it segments from the surrounding grid is complex and fraught with difficulties. DOD microgrids that rely on gas or diesel-fueled generators have suffered repeated failures when attempting to exercise their blackstart capabilities, even under conditions far more benign than would exist in a Black Sky event. The U.S. Armed Services are now conducting more numerous and realistic blackstart exercises in an effort to improve preparedness for islanded operations.[128] DOD should share the findings of these exercises with owners and operators of other large-scale microgrids, especially regarding common types of failures and mitigation options. The Department should also provide data on the design of these exercises so they can be adapted for use in microgrids that support NCFs, community lifelines, and other critical facilities.

128 See, for Example, U.S. Army Fort Riley, Black Start Exercise, https://home.army.mil/riley/about/Black-Start-Exercise

IV | BLACKSTART: A BULWARK FOR NATIONAL SECURITY IN THE UNITED STATES AND ITS GLOBAL PARTNERS

TOPs and their partners rarely get credit for the contributions that blackstart plans and capabilities can make to countering the goals that adversaries will have in attacking the grid. The electricity subsector should be recognized for the immense contributions its members make to national security. The analysis that follows examines how these contributions can help address emerging threats of special significance and can also help reduce the likelihood of attacks on the grid by strengthening deterrence.

The U.S. and its security partners face Black Sky threats (especially EMP and cyberattacks) from multiple potential adversaries, including China, Russia, North Korea, and Iran. Specific scenarios that could necessitate blackstart restoration will vary accordingly. However, for U.S. national security officials, one danger has become increasingly prominent: the risk that in a conflict over

Taiwan, China will attack the grid to disrupt the flow of forces to the conflict zone, jeopardize public health and safety, and incite social chaos.

The analysis that follows examines this scenario as a "use case" that can not only help focus blackstart preparedness initiatives for a key threat, but also highlight broader implications for meeting additional threats in the years to come. This section also examines how emergency authorities, including Grid Security Emergency orders for the U.S., might be employed to help shape blackstart restoration priorities – but only if government authorities partner with utilities to plan for and exercise such orders well before they are needed.

The U.S. and its security partners face Black Sky threats (especially EMP and cyberattacks) from multiple potential adversaries, including China, Russia, North Korea, and Iran. Specific scenarios that could necessitate blackstart restoration will vary accordingly.

A. Blackstart Preparedness against Chinese Attacks

The 2022 U.S. National Defense Strategy provides the starting point to assess potential blackstart requirements against EMP or cyberattacks from China. The Strategy states that "The scope and scale of threats to the homeland have fundamentally changed. Especially important, "The PRC or Russia could use a wide array of tools in an attempt to hinder U.S. military preparations and response in a conflict, including actions aimed at undermining the will of the U.S. pubic, and to target our critical infrastructure and other systems."[129]

The 2023 Annual Threat Assessment by the Office of the Director of National Intelligence provides an updated and still more pointed warning. The

129 Department of Defense, 2022 National Defense Strategy, 5, October 2022, https://media.defense.gov/2022/Oct/27/2003103845/-1/-1/1/2022-NATIONAL-DEFENSE-STRATEGY-NPR-MDR.PDF

Assessment finds that if Beijing feared that a major conflict with the United States were imminent, it almost certainly would consider undertaking aggressive cyber operations against U.S. homeland critical infrastructure and military assets worldwide. Such a strike would be designed to deter U.S. military action by impeding U.S. decision-making, inducing societal panic, and interfering with the deployment of U.S. forces.[130] Of course, if China were to conduct such attacks against U.S. military bases and other assets both at home and abroad, many U.S. security partners in East Asia that host such assets (and would be vital for coalition operations) should expect attacks on their grids as well.

For a scenario of this sort, or an equivalent one involving Russia or other potential adversaries, countering adversary objectives would entail effective power restoration capabilities for three sets of facilities. The first set is that of U.S. defense installations, including those that would help marshal and flow forces from inside the United States to the conflict zone. TOPs blackstart plans already prioritize the restoration of power to major military bases in their service areas. Continued collaboration with DOD should go forward to ensure that all bases that support such force deployments, including small but important installations, receive such prioritized treatment in the continental U.S., Hawaii, and territories such as Guam.

The second set of assets that merit prioritized restoration is comprised of the civilian owned and operated infrastructure necessary to support that flow, including U.S. seaports and the electric infrastructure that enables them to operate. These facilities are not Defense installations. Yet, owing to their roles in deploying forces, we should treat them as equally important assets for blackstart planning purposes. DOE, DHS (including the Coast Guard), the Department of Transportation, and DOE should launch specialized electricity planning and resilience initiatives that include ports and the transmission and distribution utilities that serve them, focused on defeating adversary efforts to disrupt U.S. force deployments.

The third set constitutes community lifelines and other functions necessary to ensure societal continuity, and defeat counter adversary goals of inciting

130 Office of the Director of National Intelligence, Annual Threat Assessment of the U.S. Intelligence Community, February 6, 2023, 10, https://www.dni.gov/files/ODNI/documents/assessments/ATA-2023-Unclassified-Report.pdf

chaos and public disorder. Previous sections of this study have already addressed the importance of blackstart restoration for these assets. However, one issue involving all three types of adversary targets merits additional analysis: the contribution of blackstart capabilities to deterrence.

Deterrence

The National Defense Strategy highlights two forms of deterrence that are relevant to blackstart and can help TOPs refine their blackstart plans and exercises. The most familiar is that of deterrence by cost imposition. This form of deterrence hinges "on our ability to impose costs in excess to the perceived benefits of aggression."[131]

Military installations responsible for cost imposition cannot function without electric power. The Defense Science Board's 2017 Report on Cyber Deterrence found that grid-provided power (supplemented by backup emergency power capabilities) is essential for ensuring that U.S. forces are ready and able to respond to enemy attacks. Accordingly, "due to the centrality of electrical power generation in supporting military strike capabilities, the cyber security and resilience of electrical power deserves particular attention, and should be supported by increased DOD collaboration with the electric power subsector, the Department of Energy, the Department of Homeland Security, and other key stakeholders in grid resilience."[132] The U.S. Cyberspace Solarium Commission final report (2020) offers additional recommendations on how investments in grid resilience can strengthen deterrence.[133] Ideally, such measures will reduce the risk that TOPs will ever need to conduct post-cyberattack blackstart operations.

The deployment of advanced microgrids and improved emergency power capabilities can help meet these power needs. Nevertheless, TOPs and their

131 National Defense Strategy, 9

132 DoD, Report of the Defense Science Board Task Force on Cyber Deterrence, February 2017, p. 19, https://www.armed-services.senate.gov/imo/media/doc/DSB%20CD%20Report%202017-02-27-17_v18_Final-Cleared%20Security%20Review.pdf

133 U.S. Cyberspace Solarium Commission, Final Report (March 2020), 23-29, https://drive.google.com/file/d/1ryMCIL_dZ30QyjFqFkkf10MxlXJGT4yv/view

partners should continue to collaborate with DOD to ensure that installations responsible for cost imposition receive the very highest priority for blackstart power, lest adversaries believe they can escape punishment by taking down the grid.

Effective blackstart capabilities can also contribute to "deterrence by resilience." To deny adversaries the benefits of aggression, the National Defense Strategy calls for measures to strengthen resilience – that is, "the ability to withstand, fight through, and recover quickly from disruption. The Department will improve its ability to operate in the face of multi-domain attacks on a growing surface of vital networks and critical infrastructure, both in the homeland and in collaboration with Allies and partners at risk."[134] Partnering with the electricity subsector to strengthen blackstart plans and capabilities to support defense operations will greatly enhance such efforts. For example, the use by TOPs of the Transmission Resilience Maturity Model and its assessment tools directly contribute to deterrence by resilience.

B. Employing Emergency Authorities to Shape Blackstart Restoration

The governments of the U.S. and many partner nations have emergency authorities that could be extraordinarily useful for countering adversary attacks on the grid, and guiding restoration operations in ways that directly support national security priorities. One such instance of emergency authority that illuminates broader challenges and opportunities when employing them is the Fixing America's Surface Transportation (FAST) Act, which added a new Section 215A to the Federal Power Act (16 USC § 824 et seq) granting the Secretary of Energy, following a Presidential Declaration of a "grid security emergency", the authority to implement emergency security measures.[135]

134 National Defense Strategy, 8

135 Fixing America's Surface Transportation Act, Public Law 114-94, December 4, 2014

Three provisions that this legislation added to the Federal Power Act (FPA) are especially significant for blackstart restoration. First, the Congress required that the Secretary of Energy designate Critical Defense Facilities (CDFs) that are "critical to the Defense of the United States."[136] Prioritizing the restoration of power to CDFs during blackstart operations will be vital for U.S. security. However, DOD is undertaking a thorough reassessment of the domestic facilities and functions it requires to execute regional contingency operations and other warfighting plans. As noted above, civilian ports and supporting intermodal transportation systems are essential for executing many such plans. As the array of CDFs evolves, keeping TOPs informed of potential restoration priorities (and integrating them into broader blackstart planning) will be essential.

The second feature of the FAST ACT of significance for blackstart lies in the new emergency authorities Congress granted to the Secretary of Energy. If the president declares that a cyberattack or other attack or disaster has created a grid security emergency (GSE), the Secretary of Energy has the authority to issue GSE orders to "protect or restore" the reliability of defense critical electric infrastructure (DCEI) – that is, any infrastructure in the continental United States that services a CDF.[137] The development, issuance, and compliance with GSE orders to guide blackstart operations will pose major difficulties, especially in a contested environment with sustained enemy operations to disrupt communications systems.

A third feature of the Act adds further complexities for tailoring blackstart to meet national security objectives. In addition to DCEI, Congress also established the category of critical electric infrastructure (CEI). CEI includes systems or assets of the BES, "whether physical or virtual, the incapacity or destruction of which would negatively affect national security, economic security, public health and safety, or any combination of such matters."[138] The Act gives the Secretary the authority to issue GSE orders to protect or restore the reliability of CEI as well. In doing so, Congress vastly increased the number and diversity of systems that the Secretary could prioritize for blackstart restoration, from

136 PL 114-94, p. 1775-6

137 The FPA states that the Secretary shall designate CDFs in consultation with other appropriate Federal agencies and with the owners and operators of infrastructure that may be DCEI. PL 114-94, p. 1773

138 PL 114-94, p. 1773

CEI that supports hospitals and water systems to financial institutions and major data centers.

The authority of the Secretary to issue such guidance could be enormously valuable in grid security emergencies. But TOPs also face problems in preparing for potential GSEs to guide blackstart. Given the multiplicity of CDFs and systems that may be designated as CEI and DCEI over time, TOPs' contingency planning for prioritized load pickup will not only be increasingly complex, but also at potential risk of conflicting government-issued priorities. Efforts to avoid such conflicts must also account for each TOP's own preexisting priorities. Some of those priorities will almost surely be shared by Federal agencies – for example, safe nuclear power plant shut down. But others, including the restoration of service to specific electric-powered compression stations essential for delivering natural gas to cranking path generators, may not be recognized by government officials. The same is true of the steps needed to grow and integrate power islands until the interconnections are re-established.

Sophisticated adversary attacks (including EMP, and potentially, kinetic strikes employing hypersonic missiles or large scale, coordinated attacks employing unmanned aerial systems against critical substations and other assets) may disable particular restoration pathways.

TOPs will need wide flexibility in deciding how best to implement government-issued priorities.

The layout of the cranking path and the imperative to maintain the stability of the emerging power island can also complicate TOP compliance with GSE orders. Regardless of how important a given CDF may be from a national security perspective, grid physics and transmission system topology will "get a vote" in the order in which TOPs will ultimately pick up loads as they add generation capacity.

Moreover, sophisticated adversary attacks (including EMP, and potentially, kinetic strikes employing hypersonic missiles or large scale, coordinated attacks employing unmanned aerial systems against critical substations and other assets) may disable particular restoration pathways. TOPs will need wide flexibility in deciding how best to implement government-issued priorities; GSE orders should be structured accordingly. But the need for such flexibility extends to the government as well. As regional crises escalate, and adversaries begin to disrupt U.S. infrastructure essential for deploying and sustaining forces in the conflict zone, priorities between CDFs may rapidly shift – with significant consequences for blackstart restoration goals.

Deeper government-industry cooperation will be essential to meet these restoration challenges and prepare for the near-inevitability that in Black Sky events, the Secretary will issue GSE orders to prioritize restoration operations. These multiple and potentially conflicting restoration priorities for societal continuity and national security also suggest a further implication for future blackstart planning and strategic concepts.

DOE should coordinate transmission operators, Reliability Coordinators, and other key stakeholders to pre-plan for the use of GSE orders in ways that facilitate the execution of blackstart plans to ensure societal continuity and meet national security priorities. These organizations should also exercise the implementation of GSE orders in ways that reflect the disrupted environment in which orders would be issued and executed. DOE, TOPs, and other blackstart partners should prepare in advance for GSEs by 1) developing "template" emergency orders that account for national security and other critical/priority loads in a given TOP service area, which DOE would modify as needed to fit the circumstances of an actual emergency; 2) exercise the issuance and implementation of such orders; 3) ensure that survivable communications are available to enable the delivery of orders and subsequent messaging; and 4) address other issues of concern to these partners, including indemnification of utilities against the unintended consequences of order compliance.

3

CHAPTER THREE

ENSURING RESOURCE ADEQUACY FOR BLACKSTART IN A DECARBONIZING GRID

INTRODUCTION

The United States and many other nations are undergoing a momentous shift in their generation fleets, with equally far-reaching implications for blackstart power generation. The retirement of coal and (in some regions) gas generators confront TOPs and their partners with far-reaching challenges for maintaining the grid's reliability. In particular, as fossil retirements accelerate, risks are intensifying to both the adequacy of generation resources and the grid's ability to withstand disturbances and limit their systemic effects – including from Black Sky events.[139] Yet, advances in inverter technologies are now

139 NERC defines the reliability of the interconnected Bulk-Power System in two aspects: resource adequacy and operating reliability, which includes the ability of the Bulk-Power System "to withstand sudden disturbances" while also "avoiding uncontrolled cascading blackouts or damage to equipment." Section A examines these reliability factors and how they might be applied to blackstart restoration. NERC, Reliability Terminology, August 2013, https://www.nerc.com/AboutNERC/Documents/Terms%20AUG13.pdf

creating opportunities to mitigate these problems and develop new strategies to strengthen grid resilience.

The transformation of the U.S. generation fleet poses an equivalent mix of risks and opportunities for blackstart restoration. Already, Regional Transmission Operators (RTOs) such as PJM are proposing to allow hybrid resources (e.g., solar and wind generation paired with energy storage) to serve as "fuel assured blackstart units." PJM states that as long as such resources can meet the same stringent performance requirements as other fuel-assured units, these resources "can add significant value to a restoration plan by increasing fuel diversity among resources that are committed as Black Start Units."[140]

The transformation of the U.S. generation fleet poses an equivalent mix of risks and opportunities for blackstart restoration.

However, energizing blackstart units (after first establishing situational awareness over the status of blackstart infrastructure) constitutes only the first phase of blackstart restoration. As highlighted in Chapter I, blackstart units must then energize transmission facilities to sequentially restart generators along a cranking path, and thereby establish an initial power island within the blacked-out region. TOPs and their partners then add additional "next start" generation (balanced with the pickup of additional load) to grow those islands, maintain their stability, and integrate and synchronize them with neighboring islands until the grid is fully restored. Moreover, all of these steps must go forward in the face of the extensive damage that the precipitating Black Sky event would create.

140 PJM, PJM Interconnection, L.L.C., Docket No. ER23-1874-000 Black Start Service Revisions to Enhance Fuel Assurance, Filing to the Federal Energy Regulatory Commission, May 12, 2023, 20, https://www.pjm.com/directory/etariff/FercDockets/7345/20230512-er23-1874-000.pdf. Note that the PJM's proposal is structured to provide opportunities for the use of what it terms "intermittent" resources ("wind, solar, and run-of-river hydroelectric resources") and "hybrid" resources ("intermittent resources paired with storage"). PJM, Affidavit of Michael E Bryson on Behalf of PJM Interconnection LLC, May 12, 2023, 47, https://www.pjm.com/directory/etariff/FercDockets/7345/20230512-er23-1874-000.pdf. Section D of this study provides a detailed analysis of PJM's tariff proposal and its implications for future blackstart system planning and capabilities in greater detail.

Chapters III, IV, and V examine the diverse but closely connected initiatives that will be necessary to ensure the adequacy and operating reliability of blackstart power generation, given 1) the growing penetration of IBRs in distribution and bulk power systems; and 2) the development and deployment of these IBRs creates new options for blackstart but also reveals novel and difficult-to-resolve problems for restoration and overall grid reliability.

The starting point for this analysis: even if IBRs replace gas or diesel-fueled blackstart units as part of the grid's broader transformation, and can successfully initiate the restoration process, the problem remains of ensuring the availability (and viability) of the subsequent cranking path and next start generation. Natural gas-fired resources provide the vast majority of these follow-on generation assets for virtually every TOP in the United States and their equivalent organizations in allied nations.

Even if IBRs replace gas or diesel-fueled blackstart units as part of the grid's broader transformation, and can successfully initiate the restoration process, the problem remains of ensuring the availability (and viability) of the subsequent cranking path and next start generation.

Yet, while natural gas-fired generation is essential for blackstart, that source of power also faces intensifying threats. The U.S. intelligence community assesses that China has sophisticated capabilities to disrupt the flow of natural gas on which such generation depends, and that Chinese actors are conducting sustained intrusion campaigns to hold that infrastructure at risk of physical damage and disruption.[141] Chapter V examines threats to natural gas fuel supplies and proposes options to mitigate the impact of pipeline attacks on power restoration operations, both for blackstart units and next generation assets.

141 Office of the Director of National Intelligence, Annual Threat Assessment of the U.S. Intelligence Community, February 6, 2023, https://www.dni.gov/files/ODNI/documents/assessments/ATA-2023-Unclassified-Report.pdf; Cybersecurity and Infrastructure Security Agency (CISA), Chinese Gas Pipeline Intrusion Campaign, 2011 to 2013, July 21, 2021, https://www.cisa.gov/news-events/cybersecurity-advisories/aa21-201a

Challenges to gas-based blackstart are also emerging in the regulatory and policy realms. A growing number of federal, state and local initiatives call for the decarbonization of power production, including gas-fired generation. Many of these state efforts emphasize the need to maintain grid reliability as decarbonization goes forward. New York, for example, calls for transitioning to a "100% zero-emission power grid" while also maintaining "a completely safe and reliable power grid."[142] Federal agencies are also coordinating to protect reliability as fossil plants are retired. Most notably, as the U.S. Environmental Protection Agency (EPA) was formulating its May 2023 proposal to sharply reduce carbon emissions by coal and natural gas-fired generation, the EPA and Department of Energy (DOE) agreed to cooperate "on electric sector resource adequacy and operational reliability."[143]

Yet, none of these Federal or state initiatives directly address the dangers that decarbonization poses to blackstart. This chapter finds that over the long term, IBRs could make valuable contributions to blackstart restoration, if the significant technical and operation problems of integrating them into TOP plans can be resolved. Indeed, as IBR penetration grows, it will be increasingly important to ensure that these assets can support restoration (rather than impeding it). Chapter VII will examine whether and how small hydro plants and other advanced generation resources might contribute to blackstart restoration as well. A principal finding of that chapter is that by increasing the diversity of blackstart generation assets, the U.S. and partner nations can significantly complicate adversary attack planning and

For many years to come, gas generation (and hydropower where available) will remain irreplaceable to grow the initial power islands that cranking path operations create, and to further expand and synchronize these islands to complete nationwide restoration.

142 New York State Climate Action Council, New York's Scoping Plan, 225,
December ,225-247 ,2022 https//:climate.ny.gov/resources/scoping-plan/

143 U.S. Department of Energy and U.S. Environmental Protection Agency, Joint Memorandum on Interagency Communication and Consultation on Electric Reliability, March 9, 2023, https://www.epa.gov/system/files/documents/2023-03/DOE-EPA%20Electric%20Reliability%20MOU.pdf

execution, in contrast to relying on a potential "monoculture" of IBRs that may be vulnerable to common mode failures.

However, for many years to come, gas generation (and hydropower where available) will remain irreplaceable to grow the initial power islands that cranking path operations create, and to further expand and synchronize these islands to complete nationwide restoration.[144]

Chapters III-V propose options to go forward with carbon reductions while also strengthening the resilience of blackstart restoration. For example, within overall policies to reduce carbon emissions, regulators and their partners could create exemptions (or "carveouts") to designate and retain blackstart-designated generators, even as other forms of fossil generation are retired. Such an approach would strike a balance between the very limited carbon emissions that such generators would create, and their immense value for restoration capabilities and societal continuity. For TOPs that rely on independent power producers to provide blackstart generation assets, targeted measures for cost recovery and regulatory relief may also be needed to ensure the viability of such carveout strategies.

Over the long term, it may also become possible for IBRs to enable blackstart in a zero-carbon grid.

The key to this option lies in the deployment of grid-forming inverters that do not need to follow the voltage signals provided by synchronous generators.

In addition, these chapters explore how - and how much - IBRs can supplement the power that gas and hydro assets provide for blackstart. The BES is now what can be termed as a "hybrid electric system" – that is, it comprises both inverter-based resources and large synchronous generators that have traditionally supported grid reliability, including fossil, hydro, and nuclear power plants.[145] TOPs now have the opportunity to adopt an equivalent

144 Coal generation requires significant time to get boilers up to the temperature needed to support generation (typically 24-48 hours) from a cold boiler, and hence offers limited value for blackstart.

145 For a characterization of this hybrid system, see Yashen Lin et al, Research Roadmap on Grid-Forming Inverters, National Renewable Energy Laboratory (NREL), November 2020, iv, https://www.nrel.gov/docs/fy21osti/73476.pdf

hybrid strategy to provide electricity for blackstart restoration, and enable IBRs that follow the voltage signals provided by synchronous generators to provide power for restoration. However, incorporating these "grid following" IBRs at scale will pose multiple problems for keeping power islands stable, as well as for related problems of load pickup, maintaining the effectiveness of protection systems, and other crucial issues. The analysis below examines these problems and potential solutions to them.

Over the long term, it may also become possible for IBRs to largely - or perhaps even entirely - replace gas generators for blackstart and enable blackstart in a zero-carbon grid. The key to this option lies in the development of grid-forming inverters to control the output of IBRs and enable them to restart the grid without needing to follow the voltage signals provided by synchronous generators. NERC notes that these grid-forming inverters (GFM) have been widely researched in battery energy storage systems (BESS), wind power plants, solar photovoltaic (PV) plants, and "hybrid plants" -- i.e., plants that combine solar and wind generation assets with BESS. [Note that the term hybrid plant refers to a specific type of facility, as opposed to broader "hybrid electric systems" that combine IBR and synchronous generation resources into a power grid.] There are installed GFM projects in Hawaii, South Australia, and other locations where operators have successfully tested these grid-forming functions, including extremely fast power injection in the inertial time frame in response to frequency events, islanded operation capability without synchronous generation, and - most important for this study - capabilities for blackstart restoration.[146]

NERC cautions that "widespread understanding of GFM controls and their impact on BES performance is still in the early stages." Indeed, above and beyond the problems of using IBRs to supplement synchronous generation, relying on GFM assets alone to complete an interconnection-wide blackstart restoration process would involve monumental (and perhaps unsolvable) technical and operational difficulties. Nevertheless, NERC concludes that GFM

146 NERC, White Paper: Grid Forming Technology: Bulk Power System Reliability Considerations, December 2021, https://www.nerc.com/comm/RSTC_Reliability_Guidelines/White_Paper_Grid_Forming_Technology.pdf

technology "shows significant promise."[147]

A comprehensive study by International Council on Large Electric Systems (CIGRE) on the impacts of high IBR penetration on the electric systems reaches a more dramatic conclusion: "It can be concluded that in future it is theoretically possible to operate the system with 100% power electronic interfaced generation, i.e., without any synchronous inertia present in the system," including for GFM-based blackstart power restoration.[148] The National Renewable Energy Laboratory (NREL) and its partners have also provided a detailed Research Roadmap on GFMs and options for the phased deployment of those systems from small island grids to very large-scale deployments across the BES.[149] This chapter examines the principal impediments to making such progress, and also the potential benefits of maintaining a mix of IBR and synchronous generation (especially gas) to complicate enemy attack planning and execution, and thereby strengthen overall blackstart system resilience.

A comprehensive study by International Council on Large Electric Systems (CIGRE) on the impacts of high IBR penetration on the electric systems reaches a more dramatic conclusion:

"It can be concluded that in future it is theoretically possible to operate the system with 100% power electronic interfaced generation, i.e., without any synchronous inertia present in the system".

However, all such blackstart system components and control operations must also be resilient against EMP effects, sophisticated cyberattacks, and other Black Sky hazards. Moreover, as noted above, adversaries can also seek to

147 NERC, White Paper: Grid Forming Technology: Bulk Power System Reliability Considerations, December 2021, https://www.nerc.com/comm/RSTC_Reliability_Guidelines/White_Paper_Grid_Forming_Technology.pdf

148 International Council of Large Electric Systems (CIGRE), Impact of High Penetration of Inverter-Based Generation on System Inertia of Networks, October 2021, 110, https://electra.cigre.org/319-december-2021/technical-brochures/impact-of-high-penetration-of-inverter-based-generation-on-system-inertia-of-networks.html

149 NREL. Research Roadmap on Grid Forming Inverters, vii, https://www.nrel.gov/docs/fy21osti/73476.pdf

disrupt gas generation by attacking natural gas transmission pipelines and other infrastructure. We need to build resilience against such indirect threats to power production, as well as direct Black Sky-scale attacks on blackstart generation assets, transmission corridors, distribution systems that serve critical loads, and the communications and control systems that enable restoration operations.

As decarbonization goes forward, an additional problem will emerge even if policymakers create carveouts for blackstart assets. Downsizing the broader gas generation fleet would create economic pressures on gas system owners and operators to downsize their supporting infrastructure as well, especially as electrification drives down the demand from other sources, including manufacturing and home heating. The resulting loss of pipeline redundancy could impede plans and capabilities to maneuver around pipeline disruptions and sustain blackstart restoration in a Black Sky environment.

The oil and natural gas (ONG) subsector and its government partners are bolstering their survivability against attack and are deepening their cooperation with the electric sector to address their interdependencies and shared risks. We need additional, specialized measures that gas and electric system operators can take to strengthen blackstart resilience before and after Black Sky events occur. We also need options to address the ripple effects of decarbonization on gas infrastructure redundancy, and the potential consequences for blackstart operations when both gas and electric systems face sustained, sophisticated attacks.

The next three chapters address all of these closely related challenges and, taken together, provide a roadmap of options to meet them.

- Chapter III highlights how the components of reliability provide a useful analytic framework for assessing options to strengthen blackstart plans and capabilities for Black Sky events. That Chapter also examines emerging problems of resource adequacy for both blackstart generation and the BES as a whole and offers recommendations to help resolve them.
- Chapter IV analyzes how the shift from synchronous generation to IBRs can threaten operating reliability of the grid and the stability of blackstart power islands. In addition, this section provides an overview of how advances in inverter technology can mitigate those dangers and help IBRs contribute to blackstart restoration. Specific areas of focus:

- The use of inverter-based resources as blackstart resources (aka blackstart units) to initiate the restoration process.
- The potential benefits of deploying numerous, widely dispersed IBR units for resilience against Black Sky hazards, as well as the regulatory and financial challenges that such deployments would entail.
- Ways in which IBRs may enable the development of new blackstart plans and capabilities in the face of deep decarbonization. For the near term, the section examines options to integrate power from grid-following resources with synchronous generation, including hydro and at least some gas-fired assets. For the longer term, the section analyzes how grid forming inverters might help drive blackstart restoration, either as a supplement to limited synchronous generation or - together with hydro - in an entirely decarbonized grid.
- Chapter V examines the current dependence of blackstart on natural gas pipelines, and opportunities to strengthen the resilience of fuel flows to support power restoration. Indeed, given that restoration systems based on grid-forming inverters would likely take decades to deploy, and may ultimately prove impractical, bolstering such resilience must become a sustained focus of collaboration. Chapter V also proposes regulatory and policy options to sustain the limited but crucial availability of gas-fired generation and supporting fuel infrastructure in a heavily decarbonized grid.

I | THE RELIABILITY CONTEXT FOR BLACKSTART

Despite investment in better controls and protection systems, the frequency of power outages in the U.S. has been increasing in the last decades, as presented in Figure 3.1,[150] The fraction of customers affected yearly by outages, considering population growth by year, is increasing by 2% on average. This trend applies to both localized power distribution outages and widespread transmission outages. Reliability has been decreasing by some measures, even as resource adequacy is increasingly threatened.

NERC defines the reliability of the interconnected BES in terms of two basic and functional aspects, both of which offer useful analogs for assessing current and future blackstart systems.

150 Mouco, A., et al. A Discussion of Resilience in Power Grids, Proc. CIGRE U.S. National Committee 2023 Grid of the Future Symposium, 2023.

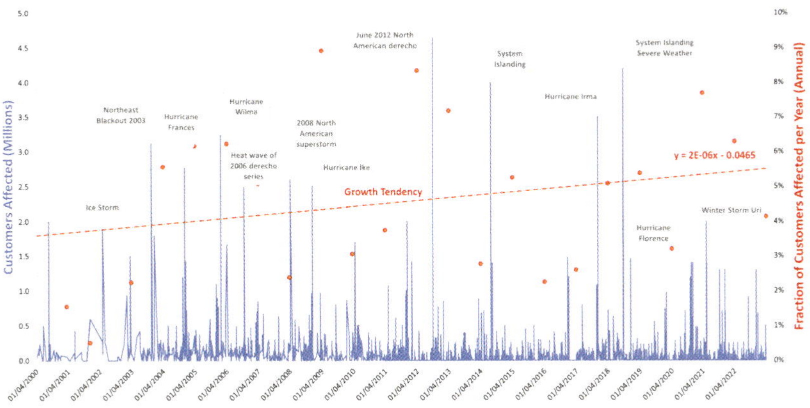

Figure 3.1 - U.S. Power Outages [T] (Reproduced with permission of the authors)

The first aspect is adequacy: that is, the ability of the electricity system to supply the aggregate electrical demand and energy requirements of the end-use customers at all times, while taking into account both scheduled and reasonably expected unscheduled outages of system elements. The second is operating reliability: i.e., ability of the BES to withstand sudden disturbances, such as electric short circuits or the unanticipated loss of system elements from credible contingencies, while avoiding uncontrolled cascading blackouts or damage to equipment.[151]

In other words, the power grid needs to achieve resilience to cascading failures in addition to adequacy and reliability. This resilience ensures that failures are kept local and small and are recovered quickly. But unlike resource

151 NERC, Reliability Terminology, August 2013, https://www.nerc.com/AboutNERC/Documents/Terms%20AUG13.pdf. PJM offers a more detailed formula to assess resource adequacy. According to PJM, "resource adequacy is the ability of the electric system to supply the aggregate energy requirements of electricity to consumers at all times, taking into account scheduled and reasonably expected unscheduled outages of generation and transmission facilities. To achieve the goal of resource adequacy, PJM maintains an Installed Reserve Margin in excess of the forecast peak load that achieves a loss-of-load expectation (LOLE) of one day in 10 years. This LOLE standard is consistent with that prescribed in the ReliabilityFirst Corporation standard for planning resource adequacy." PJM, Energy Transition in PJM: Resource Retirements, Replacements & Risks, February 24, 2023, 4, https://www.pjm.com/-/media/library/reports-notices/special-reports/2023/energy-transition-in-pjm-resource-retirements-replacements-and-risks.ashx

adequacy and operating reliability, resilience to cascading failures remains a relatively immature area of engineering for which standards have not been fully developed or implemented. A number of promising methods and standards are under development to identify and mitigate critical points of failure and increase the power grid's resilience against catastrophic failures. These include probabilistic standards that are tolerant of substantial uncertainty in the nature of the hazard that caused the blackout by emphasizing the unique role of network topology and critical assets in the interruption of cascading failures and in the recovery of the grid during blackstart.[152], [153]

Unlike resource adequacy and operating reliability, resilience to cascading failures remains a relatively immature area of engineering for which standards have not been fully developed or implemented.

Both adequacy and operating reliability are of equivalent importance for blackstart systems. For restoration to succeed, TOPs must be able to access adequate generation either from assets they own (in vertically integrated systems), or by contracting for generators and blackstart services from independent power producers (IPPs) or other external providers. Moreover, these assets must be adequate not only for initial cranking path operations, but also for the latter stages of island growth and integration that will ultimately require many thousands of megawatts of power.[154]

Operating reliability is also crucial for blackstart. As with the BES, blackstart infrastructure and operations must be able to withstand sudden disturbances and prevent disruptions at the outer edges of the restoration process from cascading back along the cranking path or from causing power island collapse.

152 Mouco, A., et al. Resilience of Power Grids to Catastrophic Cascading Failures; Proc. CIGRE U.S. National Committee 2022 Grid of the Future Symposium. 2022.

153 Mouco, A., et al. A Discussion of Resilience in Power Grids, Proc. CIGRE U.S. National Committee 2023 Grid of the Future Symposium, 2023.

154 As of February 2023, the U.S. has nearly 1.3 million megawatts of generation capacity. The largest fuel source for this capacity is natural gas (43.9%), followed by coal (17%). Wind, nuclear, and hydro together account for more than one-third of capacity. American Public Power Association, America's Generating Capacity 2023, May 2023, 1, https://www.publicpower.org/system/files/documents/Americas_Electricity_Generating_Capacity_2023_Update.pdf

Preventing such collapses will present restoration personnel with significant challenges. PJM notes that in addition to developing cranking paths from the Black Start Unit to the transmission facilities and other critical loads that must be energized to carry restoration forward, blackstart operations will typically entail other steps as well, such as bringing on load and additional generation to help stabilize the cranking path and the island it is forming. Moreover, throughout this process, "system frequencies and voltages need to be controlled to avoid tripping facilities back off-line or damaging equipment. If the restored generation is tripped, then the island formed by that generation and its associated load will be back off-line, and the restoration process will have to start over."[155]

> Operating reliability is also crucial for blackstart. As with the BES, blackstart infrastructure and operations must be able to withstand sudden disturbances and prevent disruptions at the outer edges of the restoration process from cascading back along the cranking path or from causing power island collapse.

The risks that islands will collapse and that operators will need to restart the restoration process has prompted PJM to propose the establishment of a new category of blackstart units: "Fuel Assured Black Start Units" that can run for a minimum of 16 hours without reliance on grid-supplied power." Meeting this standard will allow units to conduct multiple starts of the restoration process and address the possibility that islands formed in the early stages of this process "may trip off-line due to swinging frequency and load."[156]

Such blackstart unit capabilities are not a panacea for restart challenges. For example, multiple restarts could threaten the adequacy of battery-provided power and stored fuel supplies for backup power generation that critical substations and control centers may have, and thereby create additional problems for operating reliability. But the ability of blackstart units to initiate restarts over an extended period is valuable indeed, especially in the context

155 PJM, Black Start Service Revisions, 11-12

156 PJM, Black Start Service Revisions, 17-18

of Black Sky events.

Black Sky events will threaten the effectiveness of existing blackstart plans and capabilities in two phases. First, when EMP attacks, combined cyber-physical attacks, or other events create massive outages that necessitate blackstart, those attacks will also likely damage or disrupt restoration-essential infrastructure, control systems, communications networks, and other assets. Second, adversaries may conduct follow-on strikes to collapse emerging power islands and disrupt the restoration process, and thereby intensify pressure on U.S. and allied leaders to yield in the conflict. Capabilities for multiple restarts will be essential to help counter such risks.

When EMP attacks, combined cyber-physical attacks, or other events create massive outages that necessitate blackstart, those attacks will also likely damage or disrupt restoration-essential infrastructure, control systems, communications networks, and other assets.

Other capabilities will be vital as well. For example, as will be discussed in Chapter VIII, blackstart units and other assets will need specialized hardening measures. However, for the purposes of this chapter, the example of blackstart units illustrates a broader point: resource adequacy is tightly connected to operating reliability for both blackstart systems and the BES as a whole. These two aspects of reliability are now threatened by the broader transformation of the U.S. generation fleet, with profound implications for blackstart resilience against Black Sky hazards.

II | RESOURCE ADEQUACY

NERC's 2023 State of Reliability Technical Assessment and other, region specific RTO and ISO reliability assessments highlight the intertwined nature of adequacy and operational reliability for the emerging generation mix, for both blackstart and the bulk system. NERC's 2023 Assessment states that the BES "has never been more dependent upon the round-the-clock continuity of just in time natural gas delivery," as well as mission-critical communications services, water systems, and other sectors. NERC also emphasizes that "the providers of these goods and services have always relied upon the BES for delivery of the electrical energy they require to supply and deliver their products." This growing symbiosis across industry sectors and energy subsectors

NERC's 2023 Assessment states that the BES "has never been more dependent upon the round-the-clock continuity of just in time natural gas delivery," as well as mission-critical communications services, water systems, and other sectors.

deepens critical infrastructure interdependencies and poses "risks to BES reliability in the case of natural gas interdependency."[157]

NERC's assessment highlights two factors significant for the future of blackstart planning and operations: 1) the importance of natural gas-fired generation to overall power production and grid reliability; and 2) the interdependencies between the grid and natural gas infrastructure and operations, which will create major challenges for power when both sets of systems (and communications links between them) have been disrupted by Black Sky events.

A. The Changing Generation Fleet

The grid's continuing dependence on gas generation makes the adequacy of these gas assets and their supporting fuel delivery systems an especially significant issue. Even with the growing penetration of solar and wind generation assets, gas generation remains a key source of generation for blackstart and the grid as a whole. Gas-fired generation was the largest source—about 40%—of U.S. electricity production in 2022.[158] Moreover, while climate change is spurring Federal, state, and local initiatives to shift towards zero carbon power, severe heat events can spike the use of gas-fired generation. For example, during the heat dome that spread across much of the United States in the summer of 2023, the U.S. Energy Information Administration (EIA) assessed that record-high natural gas consumption in the United States for

Even with the growing penetration of solar and wind generation assets, gas generation remains a key source of generation for blackstart and the grid as a whole.

157 NERC, 2023 State of Reliability Technical Assessment, June 2023, 37, https://www.nerc.com/pa/RAPA/PA/Performance%20Analysis%20DL/NERC_SOR_2023_Technical_Assessment.pdf

158 U.S. Energy Information Agency USEIA), Electricity Explained, https://www.eia.gov/energyexplained/electricity/electricity-in-the-us.php

electricity generation would occur in July and August in part to meet electricity demand for air conditioning.[159]

At the same time, however, NERC emphasizes that reliably integrating inverter-based resources onto the grid is of "paramount" importance. Over 70% of the new generation in development for connecting to the BES over the next 10 years is solar, wind, and hybrid plants. NERC's assessment also notes preparing the grid to operate with increasing levels of distributed energy resources (DERs) must also be a priority in many areas. Solar photovoltaic (PV) DERs are projected to reach over 80 GW by the end of this 10-year assessment, a projected increase of 25% since NERC's 2021 reliability assessment. Moreover, a total of 12 assessment areas examined by the report are projected to double the amount of DERs in their areas by 2032.

The problem for resource adequacy: in some regions, gas-fired generators are retiring faster than IBRs are replacing their production of power. NERC notes that as new resources are introduced and older traditional generators retire, careful attention must be paid to power system and resource mix reliability attributes. For example, solar resources are not at their peak power output during the peak load periods of the year in the continental U.S. due to seasonal variations. Together with the variability of sunshine on a 24-hour basis, and the impact of cloud cover and severe weather effects, these availability issues magnify the reliability challenges posed by the loss of traditional synchronous generation that can be dispatched and available whenever needed at their stated capacity.

Within the 10-year horizon, over 88 GW of generating capacity is slated for retirement through regional transmission planning and integrated processes, including gas, nuclear, and coal-fired generation. But new IBR deployments in a number of RTO and ISO service areas are not keeping pace with these massive retirements.

Within the 10-year horizon, over 88 GW of generating capacity is slated for retirement through regional transmission planning and integrated processes,

159 EIA statistics cited in Bulk Transporter, July 17, 2023, https://www.bulktransporter.com/fleet-management/article/21269646/eia-expects-record-us-natural-gas-consumption-this-summer

including gas, nuclear, and coal-fired generation. But new IBR deployments in a number of RTO and ISO service areas are not keeping pace with these massive retirements.

The California Independent System Operator (CAISO) experienced the consequence of the resulting gap on August 14 and 15, 2020, when generation shortfalls prompted CAISO to call on utilities to initiate controlled rotating electricity outages on two occasions to maintain adequate reserves during a regional heat wave. These two load-shedding events affected 491,600 and 321,000 customers, respectively. California's electric system was ultimately unable to maintain reliable operations for the first time in almost two decades.[160]

That shortfall came in the context of the state's rapid and ongoing growth of intermittent resources like wind and solar, which have flourished, while baseload and dispatchable resources have declined. In particular, 9 GW of natural gas fired generation was removed from service within five years, including many combustion or combined-cycle plants that can respond rapidly to changing loads. Deployment of solar and wind generation assets soared at the same time, but not enough to make up for the loss of other resources.[161]

PJM is now facing the risk that an equivalent gap will emerge in its own service area, calling into question its ability to maintain its required reserve margins to manage severe weather events and other contingencies. A February 2023 report by PJM identified trends of particular concern, many of which are shared by RTOs and ISOs nationwide:

- The growth rate of electricity demand is likely to continue to increase due to electrification coupled with the proliferation of high-demand data centers in the region. PJM's long-term load forecast shows demand growth of 1.4% per year for the PJM footprint over the next 10 years. Due to the expansion of highly concentrated clusters of data centers, combined with overall electrification, certain individual zones exhibit more significant demand growth – as high as 7% annually.

160 Elliott J. Nethercutt and Chris Devon, The Intersection of Decarbonization Policy Goals and Resource Adequacy Needs: A California Case Study, March 2021, https://pubs.naruc.org/pub/55D05995-155D-0A36-315C-A161357DA070

161 Ibid

- PJM's analysis estimated that 40 GW of existing generation in its service area are at risk of retirement by 2030. This figure is composed of: 6 GW of 2022 deactivations, 6 GW of announced retirements, 25 GW of potential policy-driven retirements, and 3 GW of potential economic retirements. Combined, this represents 21% of PJM's current installed capacity.
- Retirements are at risk of outpacing the construction of new resources, due to a combination of industry forces, including siting and supply chains, whose long-term impacts are not fully known.
- PJM's interconnection queue is composed primarily of intermittent and limited-duration resources. Given the operating characteristics of these resources, PJM will need multiple megawatts of these resources to replace 1 MW of thermal generation.

Due to a variety of factors, the historical rate of completion for renewable projects has been only 5%. The projections in PJM's study indicate that the current pace of new entry would be insufficient to keep up with expected retirements and demand growth by 2030. The completion rate (from queue to steel in the ground) would have to increase significantly to maintain PJM's required reserve margin. Fortunately, a number of initiatives are underway to accelerate IBR deployments and (ideally) close the gap with retirements. In July 2023, for example, the Federal Energy Regulatory Commission (FERC) approved Order No. 2023 to speed up clogged interconnection processes that have left power generation and energy storage projects waiting years for

Fortunately, a number of initiatives are underway to accelerate IBR deployments and (ideally) close the gap with retirements. In July 2023, for example, the Federal Energy Regulatory Commission (FERC) approved Order No. 2023 to speed up clogged interconnection processes that have left power generation and energy storage projects waiting years for permission to connect to the grid.

permission to connect to the grid.[162]

But accelerating IBR deployments will not resolve all the resource adequacy problems the changing generation mix is creating. As PJM notes, solar and wind generation are intermittent and limited-duration resources. Given the operating characteristics of these resources, the deployment of multiple megawatts of these resources will be necessary to replace 1 MW of thermal generation. That factor will make closing the generation gap a persistent problem, not only for PJM but also for the Midcontinent Independent System Operator (MISO) and many other RTO/ISO service areas.[163]

Utilities that lack sufficient generation in their own service areas often import electricity from other regions. However, given the multi-region gaps that now exist between generation and demand, these strategies to help maintain reliability are at increasing risk. Dominion Energy reports that "all the Company's neighbors are facing the same generation challenges, meaning that importing power and energy at any time in a year will become more scarce [sic]." Dominion also emphasizes that as retirements continue and the grid becomes increasingly reliant on renewable energy resources, the "scarcity of dependable resources" will pose reliability concerns that need continued study, especially since the growth of data centers and electrification is increasing demand much more quickly than Dominion had previously estimated.[164] These factors will further complicate reliance on accelerated IBR deployments and power imports to bolster resource adequacy, both for overall grid reliability and for blackstart requirements for next generation and subsequent restoration phases.

---

162 FERC, Fact Sheet | Improvements to Generator Interconnection Procedures and Agreements, July 27, 2023, https://www.ferc.gov/news-events/news/fact-sheet-improvements-generator-interconnection-procedures-and-agreements; "FERC issues rule to speed grid connections for storage, renewables, other generators amid 2-TW backlog", Utility Dive, July 2023, https://www.utilitydive.com/news/ferc-interconnection-rule-queue-renewable-energy/689289/. The full proposed rule is at 184 FERC ¶ 61,054, https://www.ferc.gov/media/e-1-order-2023-rm22-14-000. The Final Rule has not yet been published in the Federal Register.

163 See NERC's assessment of these "high risk areas" for resource adequacy and reserve margins in NERC, Long Term Reliability Assessment, December 2022, 5, https://www.nerc.com/pa/RAPA/ra/Reliability%20Assessments%20DL/NERC_LTRA_2022.pdf; for a detailed account of the challenges faced by MISO, see MISO, Regional Resource Assessment (RRA), November 2022, https://cdn.misoenergy.org/2022%20Regional%20Resource%20Assessment%20Report627163.pdf

164 Dominion Energy, 2023 Integrated Resource Plan, May 1, 2023. https://cdn-dominionenergy-prd-001.azureedge.net/-/media/pdfs/global/company/2023-va-integrated-resource-plan.pdf

Moreover, above and beyond intermittency, other characteristics of IBRs will make it difficult to replace gas generation with IBRs while still keeping the grid reliable. Gas generation is well-suited to provide essential reliability services (ERS) that can help system operators respond to disturbances and prevent cascading failures. IBRs are only beginning to acquire these performance attributes. Yet, while ERS will be vital for managing severe weather events and other frequent types of incidents, Black Sky hazards are likely to overwhelm such response operations and necessitate blackstart restoration.

The transformation of the generation fleet raises an immediate question for the viability of blackstart plans and capabilities: will TOPs have the gas generators they need to conduct such restoration operations? Ensuring the adequacy of gas-fired blackstart units (including those capable of multiple restarts of the restoration process) comprises only a small part of this issue. The much larger problem lies in the availability of multi-megawatt generators essential for growing initial power islands, expanding them to serve priority loads, and integrating them with neighboring systems.

B. Blackstart Resource Adequacy

There is no publicly available data to assess whether (and to what extent) the grid-wide retirements of gas generation in general are creating specific risks to the adequacy of 1) blackstart resources (aka blackstart units) that initiate the restoration process, and 2) next start and follow-on generation. That lack of accessible information is a good thing. Given the importance of blackstart to national security, we must assume that China will specifically target blackstart infrastructure for disruption. We should continue to protect sensitive data related to blackstart and prevent the PLA from acquiring detailed, TOP-specific information concerning blackstart plans and assets.

Yet, there are ample reasons to be concerned about the adequacy of gas-fired generation for blackstart. For blackstart resources, NERC raised significant adequacy issues in the supplemental material it provided in conduction with

CIP-002-6, Cyber Security: BES Cyber System Categorization.[165] In discussions conducted during the review of that standard, concerns were raised that entities owning blackstart resources and cranking paths were removing those services in order to avoid the higher costs of complying with CIP standards. Most significant, NERC's Operating and Planning Committees have found that there has already been a reduction in Blackstart Resources because of increased CIP compliance costs, environmental rules, and other risks. Increased costs of complying with CIP standards could "result in further reduction of a vulnerable pool."[166]

NERC has reduced these compliance costs by rating blackstart resources and cranking paths as "low impact" systems: that is having low impact in terms of "the loss, compromise, or misuse" of those BES cyber systems on the reliable operation of the BES.[167] Whether that reclassification is prudent given the current security situation will be examined later in this section. However, NERC's broader point remains: blackstart resource availability is under pressure from environmental regulations, compliance costs, and other risks. An additional factor not mentioned by NERC but potentially significant: the risks of enforcement actions and heavy penalties for non-compliance with CIP standards.

NERC's Operating and Planning Committees have found that there has already been a reduction in Blackstart Resources because of increased CIP compliance costs, environmental rules, and other risks. Increased costs of complying with CIP standards could "result in further reduction of a vulnerable pool."

Next start generation faces challenges as well. A senior TOP leader who asked to remain anonymous told the Handbook authors that "tier two" (i.e., next start) gas generators are rapidly retiring in his service area. One reason is that "they are not economic – they are rarely dispatched because power from renewables is now so much cheaper."

165 NERC, CIP-002-6, Cyber Security: BES Cyber System Categorization, Supplemental Material, https://www.nerc.com/pa/Stand/Reliability%20Standards/CIP-002-6.pdf

166 NERC, CIP –002-6, 35

167 NERC, CIP –002-6, 1

Another reason is that regulators and legislators view gas generators as "dirty," and state goals for decarbonization are already driving retirements well ahead of target dates by which that process is supposed to be completed. Such anecdotal evidence is far from conclusive that next start generation is in jeopardy. However, both factors cited by this source reflect widespread issues that merit careful analysis and – in a way that protects sensitive information – comprehensive data gathering by NERC or some other appropriate organization.

The analysis that follows examines emerging risks to adequacy and options to mitigate them. This section also offers a broader framework to assess future generation requirements – especially in terms of differentiating Blackstart Resources as defined by NERC – versus the much broader generation requirements needed to complete blackstart restoration and reestablish the three U.S. interconnections.

1. Baselining Adequacy Assessments

The starting point for any such analysis is provided by the 2018 Report on Blackstart Resources Availability (BRAv), which was conducted pursuant to the 2016 FERC-NERC-Regional Entity Joint Review of Restoration and Recovery Plans and designed to provide a comprehensive understanding of the electric utility industry's bulk power system recovery and restoration planning.[168] The RCs and other participants in the Resource Availability study found that "Although several of the participants indicated that they have seen a decrease in the total number of available blackstart-capable units over time in their respective footprints, the joint study team found that participants have sufficient blackstart resources in their current system restoration plans. In addition, the joint study team found that most of the 12 participants have access to other blackstart-capable units in their footprint beyond the blackstart resources identified in their current restoration plans."[169]

168 Report on the FERC-NERC-Regional Entity Joint Review of Restoration and Recovery Plans, Blackstart Resources Availability (BRAv), May 2018, downloaded from https://www.ourenergypolicy.org/wp-content/uploads/2018/05/bsr-report.pdf; Report on the FERC-NERC-Regional Entity Joint Review of Restoration and Recovery Plans - Restoration and Recovery Report, January 2016, https://www.ferc.gov/legal/staff-reports/2016/01-29-16-FERC-NERC-Report.pdf.

169 FERC, Blackstart Resources, op cit, 11.

While many blackstart-related changes have occurred since the 2018 issuance of the report, the Resource Availability report identified trends that remain relevant today. One is the overall decline in such availability. The report found that "several participants indicated that the total number of available blackstart-capable units in their respective footprints has decreased over time. Some participants indicated that they have seen a reduction in the availability of blackstart generating units, next start generating units, or automatic load rejection units due to the impact of regulations and due to the retirement of non-economic aging assets."[170] The study also identified potential challenges in ensuring that the compensation for generation asset owners was sufficient to convince them to continue providing such services.[171]

All such trends, including regulatory and asset retirement issues, now pose increasing dangers to blackstart generation adequacy. FERC, NERC, and their partners should consider conducting a follow-on study of such adequacy. In doing so, a foundational issue will be that of the changing resource mix for blackstart. A 2022 survey of blackstart data conducted by the National Association of State Energy Officials (NASEO) found that in the United States, gas turbines represent 60 percent of blackstart units registered with the North American Electric Reliability Corporation (NERC). Hydropower units comprise 37 percent of blackstart resources and fossil fuel (combined cycle and diesel) units each are 1 percent of NERC registered blackstart units. Regional differences exist. In Texas for example, ERCOT's blackstart capabilities are comprised of 28 natural gas units at 13 sites and 13 of the units can be switched to be powered by oil if natural gas is unavailable.[172]

However, hydropower availability is at growing risk. Since 2011, annual hydropower generation in the United States has declined from over 300 billion kWh to 250 billion kWh in 2021, a 20-year low. Hydropower production in the western United States increased slightly during the 2022 due to increased

170 FERC, Blackstart Resources, 13

171 FERC, Blackstart Resources, 3-4

172 National Associate of State Energy Officials (NASEO), The Black Box of Blackstart: Electricity Reliability and Interdependency Considerations for State Energy Officials, January 7, 2022, https://www.naseo.org/data/sites/1/documents/publications/NASEO_The%20Black%20Box%20of%20Blackstart_FINAL.pdf

rainfall and snowpack, rising 13% to reach 161 million megawatt hours (MWh).[173] Nevertheless, long-term drought effects have forced the Edward Hyatt Power Plant and other hydro resources to temporarily suspend their operations.[174] Even the temporary loss of availability of such assets for blackstart could be highly disruptive for restoration. As climate change jeopardizes a growing number of hydro assets, alternative sources of power will be essential for blackstart. An updated version of the Resource Adequacy study should take such factors into account, and the possibility that gas generation will become increasingly important as a substitute for hydro, even as regulatory pressures and other factors put the future of such fossil-fueled assets in question.

2. Regulatory and Policy Challenges

PJM found in its 2023 report that "an analysis of federal and state policies and regulations with direct impacts on generation in the PJM region yielded the largest group of potential future retirements in this study." In fact, "the combined requirements of these regulations and their coincident compliance periods have the potential to result in a significant amount of generation retirements within a condensed time frame."[175] Other RTOs and ISOs have reached similar conclusions.

Twenty-two states have already enacted goals to achieve 100% carbon-free electric generation

State targets to achieve zero-carbon power generation constitute a growing risk to the reliance of TOPs on gas generators to execute their blackstart plans. Twenty-two states have already enacted goals to achieve 100% carbon-free electric generation, including California (by 2045); New York (by 2040) and Illinois (2045).[176] Climate change will almost certainly prompt other states to adopt similar targets. None of these states have exempted blackstart generators

173 EIA, After dropping to a 20-year low, western U.S. hydropower generation rose 13% last year, February 22, 2023, https://www.eia.gov/todayinenergy/detail.php?id=55599#

174 NASEO, Black Box, 9.

175 PJM, 7

176 Clean Energy States Alliance, 100% Clean Energy States, https://www.cesa.org/projects/100-clean-energy-collaborative/guide/map-and-timelines-of-100-clean-energy-states/

from their bans on fossil fueled generation. However, many of them have provisions that could open the door to consideration of blackstart-related needs.

The New York State Climate Leadership & Community Protection act, signed into law on July 18, 2019, offers a case in point. The Act sets targets to reduce emissions to 40% below 1990 levels by 2030, and then to 85% below 1990 levels by 2050. Offset measures will provide for the remaining 15% reduction of emissions to reach net zero emissions. The New York Scoping Plan that guides the implementation of the Act is highly detailed, with over 400 pages of specific actions, key strategies, and considerations for future analysis for multiple infrastructure sectors.

The implementation plan for the electricity sector is especially comprehensive and includes specific measures to "transform power generation." That section includes strategies to guide the retirement and/or repurposing of fossil fuel fired facilities and accelerate the growth of large-scale renewable generation and DERs. The plan also calls for measures by the New York Independent System Operator (NYISO) and state utilities to "enhance the grid," including upgrades in transmission and distribution infrastructure upgrades. In addition, the plan includes measures to "provide the continuity of a reliable power system" as fossil generation is replaced by intermittent zero-carbon sources of power.[177] However, the plan does not say a word about blackstart restoration. The same is true of all the other state laws and implementation guidelines reviewed for this Handbook.

It is possible that the omission of blackstart issues reflects a broader lack of awareness of their importance by state legislators and energy officials. NASEO's report on The Black Box of Blackstart constitutes an important step forward in raising that awareness.[178] As will be discussed below, the report also offers concrete recommendations to further engage state energy officials, as well as legislators and state regulators, on the continuing need for resilient blackstart capabilities.

But the potential consequences of state bans on gas-fired generation are clear. As currently framed, the New York act and its implementing guidelines

177 New York State Climate Action Council, New York's Scoping Plan, December 2022, 225-247 , https://climate.ny.gov/resources/scoping-plan/

178 NASEO, Black Box, op cit

would require the retirement of all such gas generators on which New York TOPs rely to grow and help maintain the stability of the power islands that hydropower, battery energy storage systems, or other zero-carbon energy sources (still allowed by the Act) can help them establish.

The plan does leave the door open, however, to future consideration of blackstart requirements. The document's strategy calls for "continual evaluation" of reliability-related issues, including (potentially) power restoration. In particular, the Plan specifies that:

The State should conduct established biennial evaluations to assess the state of bulk power system reliability in consultation with the federally designated electric bulk system operator (NYISO) and the New York State Reliability Council. These evaluations should ascertain if any program adjustments are needed to ensure continued safe and adequate electric service. They should be informed by the review of the State's power system performance in conformance with established operations requirements and by relevant studies including the NYISO's Reliability Needs Assessment and the NYISO 2021-2040 System & Resource Outlook.[179]

NYISO and its industry partners should consider identifying blackstart generation requirements as a key factor to consider in the State's biennial evolutions of BES reliability. As noted above, it will be essential to protect sensitive, exploitable information as any such review process goes forward.

Other states have similar exception clauses built into their zero-carbon statutes. Virgina, for example, has established the policy of developing "energy resources necessary to produce 30 percent of Virginia's electricity from renewable energy sources by 2030 and 100 percent of Virginia's electricity from carbon-free sources by 2040."[180] But the state subsequently enacted a provision that enables eligible utilities to petition the Virginia Public Utilities Commission "for relief from the requirements of this subsection on the basis that the requirement would threaten the reliability or security of electric service to customers."[181] Blackstart restoration is crucial for secure electric service.

179 Scoping Plan, 246

180 Code of Virginia, § 45.2-1706.1. Commonwealth Clean Energy Policy , 1, https://law.lis.virginia.gov/vacode/title45.2/chapter17/section45.2-1706.1/

181 Code of Virginia, § 56-585.5. Generation of electricity from renewable and zero carbon sources. (2023

Accordingly, for Virgina and other states that have equivalent exceptions to their zero-carbon goals, regulators and the industry should explore how such provisions can be applied to preserve adequate gas-fired blackstart resources and next start generation assets – unless and until IBRs can reliability provide such services.

3. Accounting for Multi-Phase Blackstart Generation Requirements

As in the 2018 FERC-NERC-Regional Entity report, it will be essential for follow-on studies to account for the full range of assets that will be necessary to conduct blackstart restoration from start to finish. NERC Standard EOP-005-3 – System Restoration from Blackstart Resources, establishes core requirements for generation and related capabilities. According to that Standard, Each Transmission Operator shall develop and implement a restoration plan approved by its Reliability Coordinator. The restoration plan shall be implemented to restore the Transmission Operator's System following a Disturbance in which one or more areas of the Bulk Electric System (BES) shuts down and the use of Blackstart Resources is required to restore the shutdown area to a state whereby the choice of the next Load to be restored is not driven by the need to control frequency or voltage regardless of whether the Blackstart Resource is located within the Transmission Operator's System.[182]

This standard also identifies a number of generator-specific requirements that can help ensure resource adequacy (along with testing, exercises, and other factors necessary for system effectiveness). Requirements include:

R1.4. Identification of each Blackstart Resource and its characteristics including but not limited to the following: the name of the Blackstart Resource, location, megawatt and megavar capacity, and type of unit.

R1.5. Identification of Cranking Paths and initial switching requirements between each Blackstart Resource and the unit(s) to be started.

R11. Each Transmission Operator and each Generator Operator with a Blackstart Resource shall have written Blackstart Resource Agreements or mutually agreed upon procedures or protocols, specifying the terms and

---

updated section) 2, https://law.lis.virginia.gov/vacodeupdates/title56/section56-585.5/

182 NERC, Standard EOP-005-3 – System Restoration from Blackstart Resources, https://www.nerc.com/pa/Stand/Reliability%20Standards/EOP-005-3.pdf

conditions of their arrangement. Such Agreements shall include references to the Blackstart Resource testing requirements.

R13. Each Generator Operator with a Blackstart Resource shall notify its Transmission Operator of any known changes to the capabilities of that Blackstart Resource affecting the ability to meet the Transmission Operator's restoration plan within 24 hours following such change.

Steven Naumann (who served on the technical expert advisory board for drafting this Handbook) notes that NERC's standards specify requirements for Blackstart Resources, such as small diesel, gas-turbines, or hydroelectric plants, which can be started without relying on external power sources. However, "these small generators do not have the capability to restore even limited portions of the electric system – their function is to provide electricity to start larger generators. Only then can system operators start restoring customer loads."[183] Consequently, even if Blackstart Resources are adequate and function as designed, restoration will fail unless the larger and more numerous generators are available to grow and integrate power islands.

Other NERC standards provide a framework to assess requirements for "next start" generation assets and other sources of power necessary to complete the restoration process. Standard EOP-006-3, System Restoration Coordination, requires reliability coordinators to "Ensure plans are established and personnel are prepared to enable effective coordination of the System restoration process to ensure reliability is maintained during restoration and priority is placed on restoring the Interconnection."[184] Standard EOP-007-0, Establish, Maintain, and Document a Regional Blackstart Capability Plan requires the Regional Reliability Organization to establish and maintain a system blackstart capability plan (BCP), as part of an overall coordinated Regional System Restoration Plan (SRP). In particular, these plans must include:

- requirements for verification through analysis how system blackstart generating units shall perform their intended functions and shall be sufficient to meet SRP expectations;

183 Steven T. Naumann, "'Black Start" Capability is Critical to a Resilient Electric Grid," Protect our Power, August 122, 2021, https://protectourpower.org/blog/black-start-capability-resilient-electric-grid/

184 NERC, EOP-006-3 – System Restoration Coordination, https://www.nerc.com/pa/Stand/Reliability%20Standards/EOP-006-3.pdf

- a database that contains all blackstart generators designated for use in an SRP within the respective areas and shall be updated on an annual basis; and
- a requirement to demonstrate that blackstart units perform their intended functions as required in the Regional SRP. This requirement can be met through either simulation or testing.

Follow-on adequacy assessments can leverage these standards and requirements to conduct a holistic assessment of generation requirements and analyze the extent to which zero-carbon regulations, economic-driven gas retirements, loss of hydro, and other factors will put overall restoration plans at risk.

4. Accounting for Black Sky Disruptions of Generation Assets

The availability of adequate blackstart generation may also be compromised by the events that necessitate power restoration. In reviewing the February 2021 extreme cold weather that affected the Electric Reliability Council of Texas (ERCOT) and nearly caused a full blackout that would have required blackstart restoration, Naumann notes that public reports found that a number of the generators that were needed to blackstart the electric system were themselves out of service or did not have fuel available.[185] A number of former commissioners from the Public Utility Commission of Texas wrote that "in the February outages, it is not clear that all of ERCOT's designated black-start assets would have been available to restart our grid due to maintenance, frozen equipment, or lack of fuel. If ERCOT had actually lost the entire power system to a full blackout, these black-start units would not have been able to do the job we pay them to do. This is unacceptable."[186]

Of course, severe weather events pale in comparison to the cumulative impact that an EMP attack could have on generator control panels and other

185 Rebecca Smith, "The Texas Grid Came Close to an Even Bigger Disaster During February Freeze, Many 'black start' units, which are used to jolt failed electricity systems, weren't working," Wall Street Journal (May 27, 2021)

186 Pat Wood, et al, Never Again: How to Prevent Another Major Texas Electric Failure, the Mitchell Foundation, June 3, 2021, https://cgmf.org/blog-entry/435/REPORT-%7C-Never-Again-How-to-prevent-another-major-Texas-electricity-failure.html

vulnerable electronic equipment. The same is true of the effects resulting from a catastrophic cyberattack specifically targeted against blackstart infrastructure and control systems. In assessing resource availability, the FERC-NERC-Regional entity study did not seek to account for the possible disruption of blackstart generation assets and other infrastructure. In the current threat environment, doing so will be essential.[187]

Establishing a design basis threat (DBT) for blackstart can help enable such analysis. Naumann notes that as cyber and other threats to blackstart assets and capabilities continue to intensify, DBTs can help system owners and operators target their investments to maintain system resilience. A DBT also can provide regulators confidence that those investments are prudent, and deal with ongoing challenges in the availability of gas-fueled cranking-path generators and next-start assets. A Black Sky-oriented DBT should include the risks posed by cyberattacks, Electromagnetic Pulse (EMP) attacks, and (perhaps) severe space weather causing a Geomagnetic Disturbance (GMD).[188] As hypersonic missile threats from China and Russia continue to evolve, it might also become prudent to include the risks posed to blackstart by the large-scale, coordinated use of such weapons against control centers and other assets and infrastructure critical for restoration.

A number of former commissioners from the Public Utility Commission of Texas wrote that "in the February outages, it is not clear that all of ERCOT's designated black-start assets would have been available to restart our grid due to maintenance, frozen equipment, or lack of fuel. If ERCOT had actually lost the entire power system to a full blackout, these black-start units would not have been able to do the job we pay them to do. This is unacceptable.

187 As will be discussed in Chapter VII, assessments of adequacy should also account for the availability of plans, capabilities, and secured storage necessary to replace damaged equipment as part of an overall blackstart resilience strategy for Black Sky events.

188 Steven T, Naumann, "Design Basis Threat Development is Key to Proper "Blackstart" Grid Capability," Protect Our Power, November 4, 2021, https://protectourpower.org/blog/design-basis-threat-development/#_ftn7

A related line of effort should examine the degree to which such criticality assessments are made. Blackstart resources and facilities on cranking paths are low-impact facilities under NERC Reliability Standard CIP-002, and thus subject to lesser requirements than other facilities. Furthermore, blackstart resources and elements of the cranking path are not covered under NERC Reliability Standard CIP-014 (for physical security) unless the substation in question is covered for other reasons.[189]

- **Recommendation:** NERC and its partners should consider whether the blackstart assets should be recategorized as medium or even high-impact under CIP-002. However, any such move could increase the regulatory burden and compliance costs for generation owners and exacerbate the trend of blackstart generation retirements that was already underway when the 2018 availability study was published.

5. Reversing Economic and Regulatory Disincentives for Generation Availability

No publicly available data exists to assess whether, on a nationwide basis, generation owners are withdrawing their assets from blackstart contracts because of 1) lack of adequate compensation, or 2) excessive regulatory risks (including enforcement actions) relative the revenue that owners can derive by providing blackstart services, as distinct from the regulatory issues associated with decarbonization. However, anonymous interviews with TOP personnel conducted by the Handbook authors suggest that both factors could help increase problems for resource availability in the future.

- **Recommendation:** A recommendation from the 2018 availability report continues to have great merit. The study suggested that ISOs, RTOs, and their partners "consider further study of the adequacy of compensation for blackstart and other resources supporting system restoration," including "any potential threat or impact on blackstart resource procurement and retention under current compensation mechanisms." The report also recommended "examining appropriate compensation for next start generating units used in system restoration,

189 Naumann, Design Basis

including compensation for costs associated with participating in blackstart related services such as expanded testing."[190] All of these recommendations should go forward as follow-on studies of adequacy are conducted, and should account for a harsh reality: compensation is a large factor in units looking to start or continue to provide blackstart service. Moreover, customers and states often do not recognize the value of having sufficient blackstart resources since, thankfully, restoration events are extremely rare. Raising their awareness of Black Sky hazards must be part of future strategies for resource adequacy.

6. Beyond Incentives: Developing Backstops for Resource Adequacy and Performance

Fewer and fewer utilities are vertically integrated these days. As a result, more and more generation is moving into a market driven construct. While financial incentives for such merchant generation is provided by entities such as PJM, these incentives are not significant enough to ensure an effective multifaceted blackstart program consistent across the U.S.. Even in the construct where a vertically integrated utility owns their generation, there are still situations where generation units critical to the black start restoration process (i.e., getting offsite power to nuclear facilities) are owned by a third parties, versus the utility responsible for the blackstart program. These third-party units may not have the same level of commitment to ensure resource availability in a crisis. In fact, some of these small units may be reluctant to connect to the grid in the early part of the restoration process, because they may be concerned that grid instabilities could damage their units. But that is the very time those units need to connect to deliver offsite power to nuclear stations and begin the restoration process.

DOE, public officials, and regulators should develop mechanisms to ensure that non-utility generators meet their obligations to provide blackstart services, regardless of extenuating circumstances.

190 FERC, Resource Availability, 3-4

- **Recommendation:** DOE, public officials, and regulators should develop mechanisms to ensure that non-utility generators meet their obligations to provide blackstart services, regardless of extenuating circumstances. One option would be for state governors to pre-plan to exercise emergency authorities that require a GO to assist the blackstart unit to restart, accompanied by liability waivers as appropriate. At the Federal level, since a black sky event will no doubt constitute a presidentially-declared Grid Security Emergency, per Sec. 215(A) (a) of the Federal Power Act, the Secretary of Energy could issue a Federal, nationwide emergency order requiring a GO to assist the blackstart unit to restart, and providing to that GO a liability waiver for such blackstart purposes. Sec 215 also provides for waiver of enforcement of regulations and enables cost recovery.[191] Over the longer terms, regulations should be established to require and enforce that GOs actually provide the blackstart services that they have contracted to provide.

7. Strengthening Collaboration

Deeper partnerships between a broad range of state and federal regulators, government officials, and industry entities (especially TOPs, RTOs/ISOs, and RCs) will be needed to ensure the adequacy of generation assets for blackstart, and balance those needs against decarbonization goals and other priorities. The NASEO "Black Box" study offers a number of partnership recommendations for state energy officials, who provide direction and planning guidance to state energy policy and programs through State Energy Plans and State Energy Security Plans.

- **Recommendation:** State Energy Plans and State Energy Security Plans should make blackstart an increasingly important area of focus to help manage risks to resource availability. Such partnerships should also be broadened to include additional blackstart resilience stakeholders.

191 DOE, "Grid Security Emergency Orders: Procedures for Issuance, "Federal Register, January 10, 2018, https://www.federalregister.gov/documents/2018/01/10/2018-00259/grid-security-emergency-orders-procedures-for-issuance#:~:text=Grid%20security%20emergency%20means%20the,or%20determination%20identifying%20the%20emergency

Exercises offer an especially valuable tool for raising awareness and consensus building for resource adequacy and emergency planning. Given the importance of blackstart restoration to societal continuity and national security, blackstart exercises should be expanded far beyond those required by NERC standards. NERC's GridEx Exercises include a wide range of industry and government participants to explore challenging restoration issues and related topics. GridEx does not, however, specifically address blackstart or the resource availability topics that are now becoming so urgent.

- **Recommendation:** Future Federal and state exercises should highlight the importance of blackstart and the imperative for adequate generation assets. Such exercises could also address challenges of prioritized load restoration to water systems and other facilities and functions, together with other operational questions related to emergency management in catastrophic, Black Sky events.

State Energy Plans and State Energy Security Plans should make blackstart an increasingly important area of focus to help manage risks to resource availability.

4

CHAPTER FOUR

OPERATING RELIABILITY FOR RESTORATION SYSTEMS IN AN IBR-HEAVY GRID

INTRODUCTION

NERC's definition of reliability includes not only adequacy but also operating reliability – that is, the ability of the BES to withstand sudden disturbances, such as electric short circuits or the unanticipated loss of system elements from credible contingencies, while avoiding uncontrolled cascading blackouts or damage to equipment.[192] As with adequacy, the ongoing revolution in the generation mix is creating challenges for the operating reliability of the BES as a whole. And again, as with adequacy, the transformation of the generation fleet will create especially severe challenges to the operating reliability of blackstart systems across the United States and in partner nations.

192 NERC, Reliability Terminology, August 2013, https://www.nerc.com/AboutNERC/Documents/Terms%20AUG13.pdf

I | IBR PERFORMANCE DEFICIENCIES: IMPLICATIONS FOR BPS RELIABILITY AND POWER RESTORATION

NERC has determined that "the rapid interconnection of bulk electric system (BES)-connected inverter-based resources (IBR) is the most significant driver of grid transformation and poses a high risk to BES reliability."[193] The most immediate risk stems from the systemic deficiencies of the inverters and other power electronics that control the flow of electricity from solar photovoltaic assets. The 2016 Blue Cut fire in Southern California demonstrated that inverters deployed at the time were incapable of riding through the transients generated by faults in the power system, and instead "tripped off" with the resulting loss

193 NERC, BPS-Connected Inverter-Based Resource Strategy, June 2022 (IBR Strategy), 1, https://www.nerc.com/comm/Documents/NERC_IBR_Strategy.pdf

of 1,200 MW of much-needed solar power.[194] Since that fire, multiple events have occurred in which momentary cessation, unwarranted inverter or plant-level tripping issues, controller interactions and instabilities, and other critical IBR performance failures have occurred.[195]

Many of these deficiencies will be remedied for future inverters by the adoption and effective implementation of new standards for IBR performance. Institute of Electrical and Electronics Engineers (IEEE) and its partners have developed IEEE Standard 2800-2022 For Interconnection And Interoperability Of Inverter-Based Resources (IBRs) Interconnecting With Associated Transmission Electric Power Systems. This standard includes performance requirements for reliable integration of inverter-based resources into the bulk power system, including voltage and frequency ride-through and dynamic active power support under abnormal frequency conditions. [196] IEEE Standard 1547-2018 establishes many of the equivalent requirements for IBR connected to distribution systems. [197]

NERC has determined that "the rapid interconnection of bulk electric system (BES)-connected inverter-based resources (IBR) is the most significant driver of grid transformation and poses a high risk to BES reliability."

Despite the adoption of these standards, however, large-scale failures of solar generation continue to occur during grid disturbances. The April 10,

194 For a summary of these events and the impact of IBR loss on grid reliability, see NERC, Quick Reference Guide: Inverter-Based Resource Activities, March 2023, 3-5, https://www.nerc.com/pa/Documents/IBR_Quick%20Reference%20Guide.pdf; and NERC, 1,200 MW Fault Induced Solar Photovoltaic Resource Interruption Disturbance Report, June 2017, v, https://www.nerc.com/pa/rrm/ea/1200_MW_Fault_Induced_Solar_Photovoltaic_Resource_/1200_MW_Fault_Induced_Solar_Photovoltaic_Resource_Interruption_Final.pdf

195 NERC, IBR Strategy, 4

196 IEEE, Standard 2800-2022 For Interconnection And Interoperability Of Inverter-Based Resources (IBRs) Interconnecting With Associated Transmission Electric Power Systems, https://webstore.ansi.org/standards/ieee/ieee28002022?gclid=Cj0KCQjwi7GnBhDXARIsAFLvH4n6NP-8bhFPPKYc0nGD9D5HWxP3MOaadbMM-Vg4NrVVMk8hNi8uHOsaAuLxEALw_wcB

197 IEEE, IEEE Standard 1547-2018 for Interconnection and Interoperability of Distributed Energy Resources with Associated Electric Power Systems Interfaces, https://www.techstreet.com/standards/ieee-1547-2018?gclid=Cj0KCQjwi7GnBhDXARIsAFLvH4l4QEiV2GXMsd35a68gJcdMlhlg6iMBF-9jDNSvgcVg7yXkE0cfsGkaAuJaEALw_wcB&product_id=1983168

2023 Southwest Utah Disturbance exemplifies these continuing threats to grid reliability the underlying problems that could undermine blackstart preparedness as well. During that event, nine solar PV facilities failed to ride through a normally cleared fault on a 345 kV transmission circuit. This resulted in an unexpected loss of 921 MW of generation. Generation Owners (GOs) had received multiple alerts from NERC regarding the need to ensure that their inverters were capable of riding through disturbances. GOs failed to do so prior to the Utah event. [198]

That failure reflects a broad problem: the unwillingness of many GOs and inverter vendors to remedy serious, well-known performance problems with IBRs. Citing the repeated ride-through failures, NERC notes that all of the causes of tripping "have been reported in past events; however, industry is not taking proactive steps to mitigate these performance issues." To be more explicit:

GOs are often not addressing performance issues that latently exist within the existing fleet. All of the causes of abnormal performance in this event have been previously documented by NERC in past reports; however, actions were not taken either by the GOs or by the inverter original equipment manufacturers (OEM) to mitigate these known risks.[199]

NERC is conducting Project 2023-02 on Performance of IBRs to ensure GOs are analyzing and mitigating abnormal performance issues when they occur.[200] NERC is also calling for the development of a comprehensive ride-through standard to replace NERC PRC-024-3, which does not address all the causes of tripping in the Southwest Utah event.[201] Additionally, NERC's Project 2020-02 contemplates modifying PRC-024-4 to apply to Synchronous Generators and Synchronous Condensers and developing a new standard, PRC-029-1, Frequency and Voltage Ride-through Requirements for Inverter-Based Generating Resources.[202] Importantly, however, no mechanisms currently exist

198 NERC, 2023 Southwest Utah Solar Loss Disturbance Report, August 2023, iv, https://www.nerc.com/comm/RSTC_Reliability_Guidelines/NERC_2023_Southwest_UT_Disturbance_Report.pdf

199 NERC, Southwest Solar, 5

200 Project 2023-02 Analysis and Mitigation of BES Inverter-Based Resource Performance Issues; https://www.nerc.com/pa/Stand/Pages/Project-2023-02-Performance-of-IBRs.aspx

201 NERC, Southwest Solar, 6

202 Project 2020-02 Modifications to PRC-024 (Generator Ride-through); https://www.nerc.com/pa/Stand/Pages/Project_2020-02_Transmission-connected_Resources.aspx

in the U.S. to ensure that when NERC or other organizations identify a problem with distribution system-connected IBRs, the owners of those resources (including home rooftop solar assets) make the necessary firmware, hardware, or software modifications necessary to remedy those problems.

The underlying issue of inattention to major IBR threats to reliability also has broader implications. The old adage is that it takes a disaster to motivate systemic change. Yet, we continue to have repeated incidents of IBRs tripping off and causing thousands of MW of power losses, even after frequent alerts regarding the importance of remedying their failure modes. The historic August 14, 2003 Northeast blackout prompted Congress to require the establishment of mandatory reliability standards for the BES, after years of having refused to do so – despite warnings that the grid faced major reliability risks.[203] The challenge today is how to transition to an IBR-heavy grid in a way that foresees and mitigates the risks of catastrophic failures.

The historic August 14, 2003 Northeast blackout prompted Congress to require the establishment of mandatory reliability standards for the BES. The challenge today is how to transition to an IBR-heavy grid in a way that foresees and mitigates the risks of catastrophic failures.

The ability of inverters to ride through disturbances will also be crucial for their use to power blackstart restoration, especially as next start assets.[204] The initial power islands established by cranking path operations will be at risk of frequency and voltage disturbances, especially as they pick up new loads and generation. Reliably integrating IBRs into this process of island growth will be contingent on their ability to ride through disturbances that are likely to occur. Those disturbances are all the more likely if adversaries sustain their attacks after creating the initial Black Sky outage, and target blackstart assets and operations for disruption.

203 Matthew Wald, The Blackout that Exposed Flaws in the Grid, The New York Times, November 11, 2023, https://www.nytimes.com/2013/11/11/booming/the-blackout-that-exposed-the-flaws-in-the-grid.html

204 The specialized requirements for using IBRs as blackstart resources, aka blackstart units, is discussed in Section C

A. Resilience for IBRs – Including for Blackstart

DOE warns that "malicious actors are well positioned to enter DER systems.[205] Many of the same vulnerabilities will apply to IBRs connected to the BES. For example, if inverters are subject to tripping off when cyberattacks or combined physical-cyberattacks create grid instabilities, China could leverage those IRB failures to exacerbate the severity of the outages they create, especially as IBR penetration increases across the U.S. and in partner nations.

Moreover, while the new IEEE standards will help ensure that newly deployed inverters have ride-through capabilities (so long as GOs and their partners ensure that the ride-through settings are fully implemented as intended), DOE warns that adversaries will still have multiple opportunities to attack such advanced inverters. One option lies in manipulating DER ride-through and trip thresholds. The Department states that such an attack would leverage the new, required grid support functionality defined in IEEE Standard 1547-2018, where DER must have adjustable voltage and frequency ride-through and expanded trip settings. Adversaries can exploit this functionality by changing the over/underfrequency or the over/undervoltage ride-through settings of many advanced DER.[206]

IEEE 1547.3-2023, Guide for Cybersecurity of Distributed Energy Resources Interconnected with Electric Power Systems, offers recommendations for cybersecurity practices and controls to ensure secure communication of DER protocols (e.g., IEEE Std 1815, IEEE Std 2030.5, SunSpec Modbus, and IEC 61850) specified in the IEEE 1547-2018. The National Institute of Standards and Technology (NIST) guidelines for Cybersecurity for Smart Inverters and NARUC's Cybersecurity Baseline also provide voluntary guidelines to design and implement DER security measures. [add new footnote here] Going forward, the National Renewable Energy Laboratory (NREL) partnering with industry and standards-setting organizations to develop still more advanced cybersecurity

205 DOE, Cybersecurity Considerations for Distributed Energy Resources on the U.S. Electric Grid, October 2022, 16, https://www.energy.gov/sites/default/files/2022-10/Cybersecurity%20Considerations%20for%20Distributed%20Energy%20Resources%20on%20the%20U.S.%20Electric%20Grid.pdf

206 DOE, Cybersecurity Considerations, 23

initiatives.[207] The IEEE and UL are also collaborating with vendors and asset operators to develop cybersecurity standards for IBRs and their associated power electronics and communications links. As with the ability of IBRs to ride through disturbances during island formation, these resources must also be able to survive direct attacks that exploit and misoperate their advanced performance capabilities.

In addition, utilities and their partners must assess broader requirements to maintain the operating reliability of the grid as IBR penetration grows. NERC stresses the importance of taking such a forward-leaning approach, including for the use of advanced IBR technologies to strengthen the operating reliability of the future grid:

The incremental cost of adding grid-forming controls to battery energy storage systems is very low; however, newly interconnecting projects are not being installed with this technology enabled.

> The BES is undergoing a rapid transition and will need to adopt new technologies to ensure reliability in the future. Proactively integrating new technologies may be drastically more cost effective than the incurred costs of possible reliability risks if no action is taken. For example, in very high IBR penetration levels, grid-forming technology may be necessary (in conjunction with synchronous condensers and other technologies). The incremental cost of adding grid-forming controls to battery energy storage systems is very low; however, newly interconnecting projects are not being installed with this technology enabled. NERC is working with industry stakeholders to identify future BES reliability risks and develop strategies for more proactive solutions to mitigate future risks rather than waiting until those risks manifest.[208]

Enabling batteries to perform grid-forming capabilities could make them enormously helpful for blackstart. That is only one of the opportunities to

207 National Renewable Energy Laboratory (NREL),Collaboration With UL Supports Development of New Requirements for Cybersecurity and Modern Energy Systems, March 14, 2022, https://www.nrel.gov/news/program/2022/collaboration-with-ul-supports-development-of-new-requirements-for-cybersecurity.html

208 NERC, IBR Strategy, 9

integrate IBRs into blackstart operations and keep power islands stable, if IBRs themselves can operate reliably in the face of frequency and voltage disturbances as islands grow.

B. Blackstart in a Hybrid Electric System

Given the transformation of the generation fleet, plans to grow and synchronize power islands will increasingly need to account for the hybrid nature of the grid – that is, a system comprised of both conventional, synchronous generation and inverter-based resources. The operating reliability of this hybrid system will depend on successfully integrating the physical properties and control responses of both types of generation.

Achieving that integration presents major unresolved challenges. DOE states that "there is no established body of experience for operating hybrid power systems with significant amounts of inverter-based resources at the scale of today's North American interconnections.[209] That same conclusion applies to the development and execution of blackstart plans that can function within (and perhaps capitalize upon) an increasingly IBR-heavy hybrid system. A prime example: if grid-forming inverters do indeed become necessary to help maintain the reliability of an IBR-heavy grid, TOPs and their partners will need to assess how best to incorporate those capabilities alongside the declining but still valuable roles played by fast ramping gas-fired generation for operating reliability and blackstart restoration.

Grid operators will also need to address the novel reliability challenges posed by increased IBR penetration in the hybrid grid, which also offers significant implications for maintaining the stability of blackstart power islands. Dave Roop notes that these resources create significant harmonics which can overheat transformers, motors, and other equipment. Moreover, to help manage these problems, grid operators rely on software that must be updated as the

209 DOE, Powering On with Grid-Forming Inverters, January 4, 2021, https://www.energy.gov/eere/solar/articles/powering-grid-forming-inverters

generation mix changes. Harmonics of the 5th, 7th and 9th are becoming an increasingly significant problem as systems incorporate a larger percentage of IBRs. The harmonics levels on some systems are as high as the 245th.

Other emerging technical issues for conducting blackstart in an IBR-heavy grid include the need to properly protect the system during this period where the protection equipment cannot differentiate between increasing load and a fault on the system. This may entail the addition of synchronous condensers to provide sufficient system strength needed to support restart effectively. A 2023 CIGRE study also notes that the stability of IBR-dominated systems can also be affected by variations in the types of loads being served. For example, directly connected small motors are notorious for causing voltage recovery issues, whereas such issues are not anticipated in systems with electronically interfaced loads like variable speed drives. Power system dynamic performance can be adversely impacted if certain types of loads dominate a local network with significant penetration of IBRs.[210] These issues will be crucial for conducting restoration and managing load pickup in the aftermath of Black Sky events.

> Given the transformation of the generation fleet, plans to grow and synchronize power islands will increasingly need to account for both synchronous and inverter-based generation sources.
>
> The operating reliability of this hybrid system will depend on successfully integrating the physical properties and control responses of both types of generation.

The analysis that follows examines the specific reliability challenges being created by the retirement of synchronous generation and the rise of IBRs, especially for providing essential reliability services that will be critical to blackstart as well as overall grid operations. This section also analyzes how technological advances can enable IBRs to provide such services, and – in

210 Ff P Mitra, et al, Stability of Inverter-based Resource (IBR) Dominated Systems with Different Types of Local Loads, CIGRE, March 2023, https://cse.cigre.org/cse-n028/stability-of-inverter-based-resource-ibr-dominated-systems-with-different-types-of-local-loads.html

combination with at least some gas and/or hydro-provided power – for blackstart in the aftermath of Black Sky events. In addition, the section explores why multiple blackstart units and cranking paths are so crucial for resilience against such events, and how hybrid IBRs that pair advanced storage with solar/wind generation might add redundancy to blackstart systems.

II | ESSENTIAL RELIABILITY SERVICES FOR INCIDENT RESPONSE AND BLACKSTART ISLAND STABILITY

NERC defines Essential Reliability Services (ERS) as those services which are vital to the reliable operation of the grid, both under normal conditions and when significant disturbances occur (including those created that would necessitate blackstart operations).[211] The performance characteristics of natural gas-fired generators are well suited to provide these services. These same capabilities will also be crucial for maintaining the stability of the power islands that TOPs establish, grow, and integrate during blackstart restoration.

211 NERC, NERC, ERS Sufficiency Guidelines White Paper, December 2016, https://www.nerc.com/pa/RAPA/ra/Reliability%20Assessments%20DL/ERS%20Abstract%20Report%20Final.pdf;.For a more detailed analysis of ERS, NERC, ERS Framework Report, December 2015

The extent to which IBRs can provide the essential reliability services now offered by gas generation will be a key issue going forward, both for the overall electric system and for blackstart in particular.

That importance is highlighted by the difficulty that utilities face in deploying new gas-fired assets to meet ramping requirements and serve peak loads, especially when solar and wind production is limited by weather or other factors. For example, Dominion Energy has proposed to build the Chesterfield Energy Reliability Center to serve as an "always ready" generation resource that can be quickly deployed on the hottest and coldest days, and serve as backup generation when other resources are unavailable or insufficient to meet customer needs.[212] However, the proposal is opposed by environmentalists who claim that the plant would threaten the state's goal of eliminating carbon dioxide emissions.[213]

The analysis below examines how advanced battery energy storage systems (BESS) can help serve peak loads and mitigate weather effects on solar and wind production. This section also examines how BESS (and IBRs in general) can help grow and maintain the stability of power islands, and supplement synchronous generation for next start and subsequent restoration phases. The use of IBRs as black start units to initiate the restoration process is examined at the end of this section.

A. Ramping as an Essential Reliability Service

To maintain the electric system's reliability, generators providing these services must be able to increase or reduce their output to help grid operators

212 Dominion Energy, Chesterfield Energy Reliability Center, https://www.dominionenergy.com/projects-and-facilities/natural-gas-facilities/chesterfield-energy-reliability-center

213 Patrick Larsen, "Dominion's Proposed Chesterfield Gas Plant Draws Opposition, VPM News, June 28, 2023, https://www.vpm.org/news/2023-06-28/chesterfield-gas-plant-dominion-virginia-clean-economy-act-climate-change

match load and generation at all times.[214] Accordingly, ramping and balancing constitute a crucial ERS for maintaining overall grid reliability. Ramping and balancing operations will also be essential yet more challenging to conduct in blackstart restoration as TOPs sequentially bring generation assets online along the cranking path and pick up loads to maintain the power island's balance. Fast ramping capabilities will help operators manage sudden imbalances that will likely occur as TOPs pick up generation and loads, including cold loads, to grow those power islands.

Ramping and balancing constitute a crucial ERS for maintaining overall grid reliability. Ramping and balancing operations will also be essential yet more challenging to conduct in blackstart restoration as TOPs sequentially bring generation assets online along the cranking path and pick up loads to maintain the power island's balance.

Gas generation and hydropower are readily available to adjust their power outputs when grid managers need them for ramping and balancing. Gas and hydro power plants are also highly dispatchable: in response to orders from grid managers, these assets can quickly raise or lower their power output to help balance generation with load.[215] As previously noted, that dispatchability depends on having adequate fuel supplies for gas generators and sufficient water to spin hydro turbines. Severe cold events and adversary attacks (including cyber, physical, and EMP strikes) can jeopardize natural gas availability; long term droughts created by climate change can endanger water availability, especially in the U.S. southwest and arid regions around the globe. Assuming these challenges can be resolved, however,

214 NERC, Essential Reliability Services, 1, https://www.nerc.com/pa/RAPA/ra/Reliability%20Assessments%20DL/ERS%20Abstract%20Report%20Final.pdf;.For a more detailed analysis of ERS, NERC, ERS Framework Report, December 2015

215 Dispatchability is defined by the Argonne National Laboratory as "the ability of a power- producing facility to provide required amounts of power (at or below the facility's nameplate rating) on demand of the grid operator ...regardless of the time of day or weather conditions." Argonne National Laboratory, Glossary, https://solareis.anl.gov/glossacro/dsp_wordpopup.cfm?word_id=4592. In addition to the utilization of BSS, other initiatives are underway to increase the dispatchability of solar and wind power. See, for example, DOE, Dispatchable Soar Power Project, January 2018, https://www.osti.gov/biblio/1418902

gas and hydro plants are well suited for ramping and balancing.

Battery energy storage systems (BESS) can help solar and wind assets provide such essential reliability services. Power production by PV arrays and wind turbines varies with weather conditions in the regions where they are deployed, the time of day, and other factors. Batteries convert stored chemical energy to direct current (DC) electrical energy, and vice versa. Inverters then convert that power to alternating current (AC) for delivery to the power grid, and also manage the reverse process for batteries that can be charged up with power from the grid (as in the case of electric vehicles).[216] Additional power electronics and control networks (including hybrid plant controls) manage the two-way power flows between these batteries and the BES or distribution systems to which they are connected, on time scales that range from microseconds to tens of seconds to minutes.[217]

> The extent to which IBRs can provide the essential reliability services now offered by gas generation will be a key issue going forward, both for the overall electric system and for blackstart in particular.

These BESS control capabilities enable them to provide near-instantaneous power for ramping and support for balancing operations.[218] Accordingly, batteries paired with solar and wind generation could help TOPs respond to grid disruptions, and, if blackstart restoration becomes necessary, to maintain the stability of power islands as those islands pick up additional loads. Such hybrid power could contribute to restoration even if grid forming inverters are never successfully deployed. In the latter case, BESS would offer a carbon free provider of fast ramping, highly dispatchable electricity, which might be combined with the limited use of gas-fired generation to manage restoration operations (including frequency-setting) in familiar ways.

216 NERC, Performance of BESS, 3

217 NERC, Performance of BESS, 3

218 Blair, Nate, Chad Augustine, Wesley Cole, et al. 2022. Storage Futures Study: Key Learnings for the Coming Decades (Golden, CO: National Renewable Energy Laboratory) 7, https://www.nrel.gov/docs/fy22osti/81779.pdf; NREL, Grid Scale Batteries, 2

The ability of BESS and other systems to provide long duration energy storage (LDES) will make these assets especially useful for ramping and restoration. DOE categorizes LDES as providing 10 to 160 hours of power dispatch. The Department's goal of having industry deploy 460 GW of long-duration energy storage by 2050 would significantly reduce requirements to contract gas-fired peaker plants. [219] To help achieve that target, DOE is spending an initial $2.8 billion to help U.S. manufactures ramp up production of storage systems, with additional funding underway in the "Long Duration Storage Shot" to pursue advanced LDES technologies.[220]

States are funding their own LDES initiatives. In New York State, for example, the New York State Energy Research and Development Authority (NYSERDA) is administering $15 million for four demonstration projects to develop long duration energy storage solutions that harness and provide stored renewable energy to the electric grid. The awards support long duration energy storage projects, which are 10 to over 100 hours in duration at rated power. The announcement supports the goal of the Climate Leadership and Community Protection Act to install 3,000 MW of energy storage by 2030 and facilitate additional development up to 6,000 MW.[221]

The announcement supports the goal of the Climate Leadership and Community Protection Act to install 3,000 MW of energy storage by 2030 and facilitate additional development up to 6,000 MW.

219 Stephen Singer, "DOE promotes commercialization of long-duration energy storage," Utility Dive, May 9, 2023,

220 DOE, Long Duration Storage Shot, July 2021, https://www.energy.gov/sites/default/files/2021-07/Storage%20shot%20fact%20sheet_071321_%20final.pdf; Robert Walton, "DOE taps 20 companies to receive $2.8B for battery manufacturing, minerals processing build-out," Utility Dive, October 20, 2022, https://www.utilitydive.com/news/doe-battery-manufacturing-minerals-processing-award/634562/

221 "New York Governor Announces Awards worth $15 Million to Four Long Duration Energy Storage Projects," T&D World, August 14, 2023, https://www.tdworld.com/distributed-energy-resources/energy-storage/article/21272394/new-york-governor-announces-awards-worth-15-million-to-four-long-duration-energy-storage-projects?utm_source=TW+TDW+Energizing&utm_medium=email&utm_campaign=CPS230825061&o_eid=4708I6058334B1I&rdx.ident[pull]=omeda|4708I6058334B1I&oly_enc_id=4708I6058334B1I

However, there is an important distinction between IBR storage dedicated to blackstart versus for supporting "normal" peaking, ramping, and balancing operations. To be useful for blackstart, LDES must have dispatchable power always available (and held in reserve) to conduct restoration operations. These storage systems must also be able to survive the disruptive effects of the initial black sky event and possible follow-on attacks. Meeting both of these requirements will take dedicated funding and cost recovery initiatives.

B. Frequency Response, Voltage Support, and Advanced IBR Capabilities

Two other essential reliability services will be necessary for incident response and blackstart island stability. Frequency support services help maintain the 60 Hz frequency of the synchronized BES and the continuous balance between load and generation resources.[222] Voltage support services are essential to maintain the voltages in the transmission system within a secure, stable range, under pre- and post-contingency situations, and thereby prevent voltage collapse or system instability that could otherwise occur.[223]

> IEEE Standard 1547-2018 established requirements for inverter ride-through capabilities that can help enable IBRs to contribute reliable power for blackstart. This standard and its BES equivalent also enable inverter based resources to help manage voltage and frequency instabilities.

The spinning mass of turbines, rotors, and other equipment that comprise gas and hydro generators (as well as nuclear power plants) provides inertia to electric

222 NERC, Essential Reliability Task Force, Concept Paper on ERS that Characterizes BPS Reliability, October 2014, https://www.nerc.com/comm/Other/essntlrlbltysrvcstskfrcDL/ERSTF%20Concept%20Paper.pdf

223 Concept Paper, 1.

systems that helps maintain their reliability. For example, when incidents cause grid frequencies to drop, the inertia from these conventional generation assets provides for automatic fast frequency response that limits the extent of the drop. Operators can then adjust the aggregate output of these assets to restore the grid's frequency to the nominal 60 Hertz.

IBRs lack spinning mass and cannot provide system inertia. However, their inverters and power electronics can adjust the flow of power they produce to provide equivalent benefits for grid reliability – and for the stability of power islands. As noted above, IEEE Standard 1547-2018 established requirements for inverter ride-through capabilities that can help enable IBRs to contribute reliable power for blackstart. This standard and its BES equivalent also enable inverter based resources to help manage voltage and frequency instabilities.[224] In particular, the document establishes voluntary technical specifications and performance requirements for DER components and systems, including those for frequency response and voltage support when grid disturbances occur. Underwriters Laboratory (UL) and other organizations are now testing and certifying DER devices as compliant with those standards.[225]

Equivalent standards are now being applied to IBRs that are connected to the BES. IEEE Standard 2800-2022 for Interconnection and Interoperability of Inverter-Based Resources (IBRs) Interconnecting with Associated Transmission Electric Power Systems provides minimum requirements for the interconnection, capability, and lifetime performance of inverter-based

224 Institute of Electrical and Electronics Engineers (IEEE), IEEE Standard 1547-2018, Standard for the Interconnection and Interoperability of Distributed Energy Resources with Associated Electric Power Systems Interfaces, February 2018. Standard 1547-2018 is only available by purchase and cannot be reproduced in whole or part without permission of IEEE. Hence, this study draft does not quote from the Standard, but instead paraphrases some of its key findings and recommendations and uses detailed analysis of the Standard provided by other reports.

225 Most important UL 1741 Ed. 3 provides a certification standard for smart inverters that help manage power flows from solar DER and other assets. IEEE, Timeline for Rollout of IEEE Std 1547-2018 Compliant DER, May 2022, https://sagroups.ieee.org/scc21/standards/1547rev/#:~:text=Timeline%20for%20Rollout%20of%20IEEE%20Std%201547%E2%84%A2%2D2018%20Compliant%20DER&text=2%20revision%20was%20published%20September,some%20consistency%20issues%20in%20Ed. See also UL, Standards and Publications, https://standardscatalog.ul.com/ProductDetail.aspx?productId=UL1741. This standard is also available only for purchase.

resources interconnecting with transmission and sub-transmission systems.[226]

IBRs that meet these standards will be much more useful for incident response and blackstart restoration support than would otherwise have been the case. Beginning with the Blue Cut Fire in 2016, inverters have repeatedly "tripped off" and stopped providing power when wildfires, severe weather, and other non-cyber events created grid disturbances (especially in terms of the grid's frequency and voltage levels). The resulting loss of IBR-provided power significantly increased the scale and severity of those disturbances.[227] If TOPs begin relying in part on IBRs to generate power for blackstart restoration, the tendency of these resources to trip offline when instabilities occur during island formation and growth would be extraordinarily disruptive.

Given the instabilities to which initial power islands will be prone, and the risk that voltage and frequency dips and surges will cause those islands to collapse, IEEE requirements for ride through and voltage/frequency response capabilities will be a prerequisite for using IBRs to supplement power from synchronous resources to conduct blackstart.

Inverters that comply with IEEE Standards 2800-2022 and 1547-2018 must be able to ride through such disturbances. They must also provide frequency response and voltage regulation capabilities when disturbances occur.[228] Both features will help them support the grid's operating reliability. Both will also make inverter-based resources better suited to provide power for blackstart restoration. Given the instabilities to which initial power islands will be prone, and the risk that voltage and frequency dips and surges will cause those islands to collapse, IEEE requirements for ride through and voltage/frequency response capabilities will be a prerequisite for using IBRs to

226 IEEE, 2800-2022 - IEEE Standard for Interconnection and Interoperability of Inverter-Based Resources (IBRs) Interconnecting with Associated Transmission Electric Power Systems, April 2023, https://ieeexplore.ieee.org/document/9762253#citations

227 NERC, Quick Reference Guide, 3-5,

228 1547-2018 also specifies that DERs may provide inertial response – that is, the capability to modulate active power in proportion to the rate of change of frequency. David Narang, "Highlights of IEEE Standard 1547-2018," NREL, October 18, 2018, 24, 28, https://www.nrel.gov/docs/fy20osti/75436.pdf

supplement power from synchronous resources to conduct blackstart.

The availability of IEEE-compliant inverters that can play such a supplementary role is rapidly growing. Dozens of vendors are ramping up their production of such products. Moreover, on February 12, 2020, the Board of Directors of the National Association of Regulatory Utility Commissioners (NARUC) unanimously approved a resolution recommending that "State commissions, consistent with the practices and procedures of that State, convene proceedings that engage stakeholders soon; utilize existing research and experience and make evidence-based decisions to adopt the current IEEE 1547-2018; and align implementation of the standard with the availability of certified equipment."[229] These steps provide a growing base of ERS-capable resources for potential blackstart use including hybrid facilities that could offer large scale and increasingly long duration energy storage for restoration operations. [230]

Of course, few currently deployed BESS or solar and wind IBRs are compliant with IEEE 1547-2018. In the U.S. alone, millions of inverters have already been deployed that lack ride-through and voltage/frequency support capabilities. Whether these IBRs might be retrofitted in a cost-effective way to support blackstart is unclear.

Moreover, even for IBRs that are compliant, many additional problems will have to be resolved before IBRs can reliably supplement the electricity provided by synchronous generators to 1) help establish initial power islands, 2) contribute to next-start generation, and 3) help operators grow and synchronize multiple power islands in the later phases of restoration. Those difficulties will be greater still if IRBs are to entirely replace (rather than supplement) gas generators in a zero-carbon grid, as will be discussed in Chapter VI.

229 National Association of Regulatory Utility Commissioners (NARUC), Resolution Recommending State Commissions Act to Adopt and Implement Distributed Energy Resource Standard IEEE 1547-2018, February 12, 2020, https://pubs.naruc.org/pub/E86EF74B-155D-0A36-3138-B1A08D20E52B

230 Section D examines the use of such hybrid plants.

5

CHAPTER FIVE

ENSURING RESILIENT NATURAL GAS SUPPLIES FOR BLACKSTART RESTORATION

INTRODUCTION

The continued importance of gas-fired generation will make it a prime target for attack. The most obvious strategy for disrupting blackstart is to directly attack the infrastructure and supporting networks essential for such operations. However, adversaries may also attack indirectly, by disrupting the flow of natural gas on which generators depend to 1) energize cranking paths and create initial power islands; and 2) grow and integrate those islands as restoration goes forward.

Recent Russian operations against Canda's oil and gas infrastructure highlight the severity of this threat vector, especially in the context of Russia's invasion of Ukraine. Canada's Communications Security Establishment (CSE) signals intelligence agency reported in June 2023 that "we assess there is an even chance of a disruptive incident in the oil and gas sector in Canada caused by Russia-aligned actors, due to their higher tolerance for risk, the increase in their numbers and activity, as well as the number of vulnerable targets in the sector overall," it said in a threat assessment.[231]

Yet, China poses an especially significant cyber threat to natural gas pipelines and their security partners. The 2022 Annual Threat Assessment by the Office of the Director of National Intelligence warns that "China almost certainly is capable of launching cyberattacks that could disrupt critical infrastructure services within the United States, including against oil and gas pipelines."[232] Earlier ODNI assessments were still more explicit. The January 2019 version stated that: "China has the ability to launch cyberattacks that cause localized, temporary disruptive effects on critical infrastructure—such as disruption of a natural gas pipeline for days to weeks—in the United States."[233] Since that time, Chinese cyberattack capabilities have continued to improve. The Department of Defense (DoD) warned in November 2021 that China "presents a sophisticated, persistent cyber espionage and attack threat to military and infrastructures systems, designed to achieve the physical disruption of critical infrastructure" and other destructive effects.[234]

The most obvious strategy for disrupting blackstart is to directly attack the infrastructure and supporting networks essential for such operations. However, adversaries may also attack indirectly, by disrupting the flow of natural gas on which generators depend

Depending how much devastation China or other potential adversaries want to create, they could launch simultaneous attacks against pipelines across multiple U.S. regions. But the disruption of even one or two major interstate transmission pipelines for weeks could have drastic effects in many regions of the United States and in partner nations as well.

231 Reuters, Russia-aligned hackers pose threat to Canada energy sector, spy agency says, June 21, 2023, https://www.reuters.com/world/russia-aligned-hackers-pose-threat-canada-energy-sector-spy-agency-2023-06-21/

232 Office of the Director of National Intelligence, Annual Threat Assessment of the U.S. Intelligence Community, February 6, 2023, 10, https://www.dni.gov/files/ODNI/documents/assessments/ATA-2023-Unclassified-Report.pdf

233 Office of the Director of National Intelligence, Worldwide Threat Assessment of the U.S. Intelligence Community, January 29, 2019, p. 5. https://www.dni.gov/files/ODNI/documents/2019-ATA-SFR---SSCI.pdf

234 Military and Security Developments Involving the People's Republic of China, DOD report to Congress, 2021, November 3, 2021, p. 88, https://media.defense.gov/2021/Nov/03/2002885874/-1/-1/0/2021-CMPR-FINAL.PDF

I | THE STRATEGIC SUPPORT FORCE OF THE PEOPLE'S LIBERATION ARMY

As with TOPs and other components of the electricity subsector, energy pipeline companies in the Oil and Natural Gas (ONG) subsector are aggressively improving the cyber resilience of their systems.

Nevertheless, pipeline owners/operators (and TOPs) face an immense challenge -- one that makes their continued defensive improvements all the more vital. Their opponent is the People's Liberation Army's Strategic Support Force (SSF), a theater command-level organization responsible for cyberattacks against U.S. critical infrastructure and other targets, as well as for psychological warfare and additional missions. Indeed, DOD's 2021 report on the Chinese military notes that the expansion of the SSF's attack capabilities and doctrine is "consistent with China's goal to become a world-leading cyber power by 2035."[235]

235 DOD, Military Developments, pp. 64 and 79.

The SSF targets a wide array of U.S. infrastructure. Yet, its focus on natural gas pipelines is extraordinary – and extraordinarily important – from the perspective of blackstart resilience. The length and intensity of China's efforts to penetrate pipeline systems, and implant compromises for use in future crisis, is exemplified by the July 2021 alert issued by the Cybersecurity and Infrastructure Security Agency (CISA) on the Chinese Gas Pipeline Intrusion Campaign. The alert revealed that China had targeted 23 U.S. natural gas pipeline operators from 2011 to 2013 in an intrusion campaign.[236]

CISA and the FBI found that of the 23 known targeted entities, "13 were confirmed compromises, 3 were near misses, and 7 had an unknown depth of intrusion." Pipeline owners/operators have improved their cybersecurity since those successful attacks occurred.

Two findings of the alert are especially significant. First, "CISA and the FBI assess that these actors were specifically targeting U.S. pipeline infrastructure for the purpose of holding U.S. pipeline infrastructure at risk. Additionally, CISA and the FBI assess that this activity was ultimately intended to help China develop cyberattack capabilities against U.S. pipelines to physically damage pipelines or disrupt pipeline operations."[237] TOPs and their partners must assume that China will continue to prepare the battlefield for future crises, and will be ready to disrupt gas flows including those blackstart restoration relies on in order to extend and intensify the coercive effects (and impacts on U.S. defense) created by large-scale outages.

A second finding of the alert also provides a useful data point. CISA and the FBI found that of the 23 known targeted entities, "13 were confirmed compromises, 3 were near misses, and 7 had an unknown depth of intrusion."[238] Pipeline owners/operators have improved their cybersecurity since those successful attacks occurred. But the SSF is increasing the sophistication of its intrusion and attack capabilities as well and will likely seek to achieve the same (or better) penetration rates in future campaigns.

---

236 Cybersecurity and Infrastructure Security Agency (CISA), Alert (AA21-201A), Chinese Gas Pipeline Intrusion Campaign, 2011 to 2013, July 20, 2021, https://us-cert.cisa.gov/ncas/alerts/aa21-201a

237 Ibid

238 Ibid

One additional risk deserves consideration: the possibility that cyberattacks attacks could create explosions (which might occur due to gas overpressure, e.g.) that would not only disrupt gas flows, but also require extensive and potentially time-consuming repair/replacement operations. State regulators across the east coast would almost certainly order the suspension of gas operations until such time as the operators can conclusively demonstrate that their systems are clean and can safely operate. Federal Departments (and, ultimately, the National Security Council) may also help determine when pipelines should go back into operation.

ONG systems often have pneumatic mechanisms to prevent over pressurization and monitoring systems that alert to unplanned pressure changes and anomalous conditions. These are driven in part by federal pipeline safety regulations intended to help detect and mitigate mis-operation. However, as China sustains its campaign to access pipeline ICS and monitoring systems, it may seek to use that access to exploit the monitoring and control systems intended to prevent over pressurization, and mis-operate control equipment by issuing malicious commands; returning false sensor information to initiate unnecessary or dangerous actions, including: delaying or blocking the flow of information used by operators; modifying alarm thresholds or other safety settings and configurations; and other means of attack.[239]

A. Implications of the Colonial Pipeline Attack for Blackstart Fuel Resilience

The Colonial Pipeline attack on May 7, 2021, is instructive to consider, though it differs from how attacks against the flow of fuel for blackstart are likely to occur. The perpetrators of the attack were reportedly members of the DarkSide criminal syndicate in Russia.[240] While criminal gangs may

239 GAO, Actions Needed to Address Significant Weaknesses in TSA's Pipeline Security Program Management, p. 11.

240 Jeff Martin, "Gas Crunch from Cyberattack Intensifies in Nation's Capital," AP News, May 14, 2021, https://apnews.com/article/europe-technology-hacking-business-472c7e4f30649aec2dbf38200521b906

have working relationships with the Russian government, it is unlikely that disruptions of natural gas supplies for blackstart will result from ransomware plots. Those disruptions will almost certainly stem from nation state attacks motivated by the highest of geopolitical stakes. Moreover, the Colonial Pipeline carries refined gasoline and jet fuel, versus the natural gas crucial for blackstart operations.

In other respects, however, the Colonial attack offers useful threat indicators. One factor to consider is the effect that IT-based attacks can have on pipeline operations. Although the ransomware attack occurred over the internet and affected IT data sets and operations, the pipeline executives responded by quickly shutting down the flow of fuel. Colonial Pipeline CEO Joseph Blount testified to Congress that "If you think there is even a 1 percent chance that that criminal got into your [operational technology] system and could potentially take over control of a 5,500-mile pipeline moving 100 million gallons a day, then you shut that pipeline down."[241] Natural gas pipeline operators (and state regulators) may face similar incentives for shut-downs in the face of nation-state attacks, especially if adversaries seek to mask or corrupt the situational awareness of system operators.

The calculations that prompted executives to shut down the Colonial Pipeline also reflected a more specific concern. System operators couldn't run the risk that hackers might "move laterally" through the company's infrastructure and cause lasting damage. If the hackers had done so, the disruptions of gas flows could have been significantly longer.[242] That risk may apply to natural gas pipelines as well. In February 2020, CISA announced that it had responded to a cyberattack affecting the operational technology (OT) network of a natural gas system. Especially significant: "A cyber threat actor used a Spearphishing Link [T1192] to obtain initial access to the organization's information technology (IT) network before pivoting to its OT network. The threat actor then deployed commodity ransomware to Encrypt Data for Impact [T1486] on

241 Aaron Gregg, "Colonial Pipeline was Shut Down with Worst-case Scenario in Mind, Executives Say," Washington Post, June 9, 2021, https://www.washingtonpost.com/business/2021/06/09/colonial-pipeline-mandiant-house-hearing/

242 Ibid

both networks."[243] Opportunities for such crossover attacks are likely to grow as pipelines install new additional system monitoring and diagnostic software to operate pipelines more efficiently.

The Colonial Pipeline attack also highlighted the psychological dimension of infrastructure disruptions, and their potential impact for blackstart fuel deliveries. Shortly after the attack occurred, the New York Times reported that because storage tanks contained so much fuel, "the attack was unlikely to cause immediate disruptions."[244] Storage facilities for natural gas may similarly mitigate the effects of pipeline disruptions in the early stages of blackstart restoration.

Just days after the Colonial Pipeline shutdown, however, 88 percent of gas stations were out of fuel, with severe shortages in gas availability in Virginia, North Carolina, South Carolina, Georgia, and other states.[245] Panic buying drove these shortages and quickly overwhelmed stored gasoline reserves. Of course, citizens and companies cannot back their cars up to interstate natural gas pipelines and demand to be filled up. But the psychological effects of natural gas cutoffs could nevertheless be intense, including for the residents of homes heated by natural gas, and for their elected representatives.

Moreover, the ransomware gang that conducted the Colonial Pipeline attack did not accompany its demands with information operations designed to inflame public panic. The DOD report on Chinese military developments emphasizes the risk that China will pair cyberattacks on U.S. infrastructure with information operations (IOs) to shape public fears and maximize coercive pressure on U.S. leaders.[246] Combined information-cyberattacks operations could also complicate blackstart operations in future confrontations.

243 CISA, Alert (AA20-049A), Ransomware Impacting Pipeline Operations, February 18, 2020, https://us-cert.cisa.gov/ncas/alerts/aa20-049a

244 David Sanger, "Cyberattack Forces a Shutdown of a Top U.S. Pipeline Operator," New York Times, May 8, 2021

245 Jeff Martin, "Gas Crunch from Cyberattack Intensifies in Nation's Capital," AP News, May 14, 2021, https://apnews.com/article/europe-technology-hacking-business-472c7e4f30649aec2dbf38200521b906

246 DOD Report, pp. 78, 86-89

II | SPECIFIC ATTACK VECTORS

Cyberattacks on natural gas system components are a prime concern. Some pipelines that carry gas to power generators are "dated and vulnerable to malicious cyberattacks."[247] As noted above, the ODNI's 2022 Worldwide Threat Assessment warned that China has the ability to launch cyberattacks that could disrupt a natural gas pipeline "for days to weeks."[248] Scaled up to simultaneously disrupt multiple pipelines, such long duration interruption of fuel for generators would have catastrophic effects on blackstart. In addition to pipeline infrastructure, ICS, and other network-connected electronic devices and control mechanisms are embedded across gas system components, from

247 Joseph R. Dancy and Victoria A. Dancy, "Terrorism and Oil & Gas Pipeline Infrastructure: Vulnerability and Potential Liability for Cybersecurity Attacks," Oil and Gas, Natural Resources, and Energy Journal Vol. 2, No. 6 (March 2017): p. 618.

248 Coats, "Worldwide Threat Assessment of the U.S. Intelligence Community," p. 5.

production, gathering, storage, and processing facilities to local distribution company infrastructure. All such connected systems are vulnerable to a greater or lesser degree to cyberattacks.

Ironically, pipeline safety requirements mandating the installation of remote-control valves are increasing the cyberattack surface. If successful, adversaries carrying out these attacks could cause significant operational disruption in the ONG subsector, and possibly create cascading failures across dependent infrastructure sectors. For blackstart operations specifically, disruptions of fuels for power generation can disrupt restoration in two particularly troublesome ways. First, fuel supply disruptions can cripple the ability of generators along the cranking path to help establish initial power islands. Second, once initial power islands are established, larger gas-fueled generators will be essential to pick up significant load and grow power islands as restoration goes forward. Attacking gas supplies can enable adversaries to disrupt these latter phases of blackstart restoration.

For blackstart operations specifically, disruptions of fuels for power generation can disrupt restoration

Given the risk that cyberattacks or other events will disrupt gas supplies for blackstart operations, TOPs and their generation asset providers have long emphasized the resilience value of employing "dual fuel" generators.[249] That value has grown in tandem with the intensifying threats to gas systems. In addition to their primary fuel (natural gas), these dual fuel generators can use diesel, no. 2 fuel oil, or other secondary sources of fuel to sustain operations when gas flows are disrupted.[250] Many of these dual fuel blackstart assets are supported by onsite oil or gas storage sufficient to run those generators for 48 to 72 hours.[251] Those run times might be sufficient for blackstart operations in

249 ISO New England, Operational Fuel-Security Analysis, January 17, 2018, p. 52; North American Electric Reliability Corporation, 2013 Special Reliability Assessment: Accommodating an Increased Dependence on Natural Gas for Electric Power Phase II: A Vulnerability and Scenario Assessment for the North American Bulk Power System, May 2013, p. 4, https://www.nerc.com/pa/RAPA/ra/Reliability%20Assessments%20DL/NERC_PhaseII_FINAL.pdf.

250 Electric Infrastructure Security (EIS) Council, E-PRO Handbook II: Volume 1 – Fuel, 2016, p. 24.

251 Federal Energy Regulatory Commission and the North American Electric Reliability Corporation, FERC-NERC-Regional Entity Joint Review of Restoration and Recovery Plans – Recommended

an uncontested environment. However, in the face of APTs and other ongoing threats to blackstart restoration and natural gas systems in a nationwide blackout, much more secondary fuel will likely be required.

The resupply of secondary fuels to generating facilities would be essential for blackstart under these contested conditions, but also constitutes a special challenge for blackstart resilience. First, we must assume that adversaries are aware that secondary fuel supplies provide a critical hedge against gas system failures. Accordingly, we should also assume that adversaries might seek to disrupt both natural gas and secondary fuel supplies for blackstart generation. Second, demand for secondary fuels will greatly exceed supply. The enormous demand that would emerge for backup generator fuel from hospitals, water systems, and other infrastructure sectors dependent on grid-provided power would create significant competition for scarce fuel resources. Third, disruptions to transportation and communications systems will complicate delivery operations. One ISO notes that fuel delivery logistics in the face of a "withering delivery supply chain" is considered an "unquantifiable X factor" for resupply operations.[252]

Study: Blackstart Resources Availability (BRAv), May 2018, p. 22, https://www.ferc.gov/legal/staff-reports/2018/bsr-report.pdf.

252 ISO New England, Operational Fuel-Security Analysis, January 2018, pp. 14 and 16, https://www.iso-ne.com/static-assets/documents/2018/01/20180117_operational_fuel-security_analysis.pdf.

III | COLLABORATIVE MEASURES TO STRENGTHEN THE RESILIENCE OF BLACKSTART FUEL SUPPLIES

One helpful step to mitigate the effects of gas system attacks would be to ensure that utilities are promptly notified of pipeline disruptions. Utilities need immediate data on such disruptions to assess their potential impact on blackstart plans and analyze contingency options. Real-time information on linepacks and expected rates of decline will also assist adaptive planning, and will be essential for startup and/or system support for starting non-gas fueled generation. Utilities, gas companies, and their partners should assess the adequacy of existing "red

> Utilities, gas companies, and their partners should assess the adequacy of existing "red phone" notification systems and protocols to support blackstart and ensure the resilience of those systems against adversary attack.

phone" notification systems and protocols to support blackstart and ensure the resilience of those systems against adversary attack.

FERC issued a final rule under Order No. 787 that allowed interstate natural gas pipelines and electric transmission operators to share nonpublic operational information to promote the reliability and integrity of their systems. Since the inception of this rule and the subsequent incorporation of those rules into the associated tariffs followed by the appropriate confidentiality agreements, natural gas and electric entities have been able to freely share operational data.[253]

NERC notes that gas-electric communications could be especially helpful during emergencies. According to a March 2023 report, "Entities of both industries could share normal and emergency conditions in real-time and ensure that each entity understands the implications to their respective systems. This could include natural gas and electric entities proactively reaching out to the operators of stressed natural gas systems to discuss the impacts, adverse or otherwise, of their expected or available actions. Entities could understand the direct impacts to electric generation assets when natural gas pipelines are directed under more extreme natural gas system operating conditions and/or force majeure conditions."[254]

NERC's report does not address the applicability of such efforts to blackstart, but the benefits of doing so could be significant, and should become a focus of future collaborative discussion between the two subsectors. Utilities could also ensure that natural gas infrastructure is not located on electrical circuits that are slated for manual or automatic underfrequency load shedding in emergencies.[255]

Network interdependency analysis can identify natural gas network facilities and lines that are especially critical for a resilient power grid, so that these most critical natural gas system assets can be prioritized for hardening investments and for operational recovery.

Measures taken in the aftermath of the Colonial Pipeline incident also provide a foundation for progress. That incident involved the flow of liquid

253 NERC, Natural Gas and Electrical Operational Coordination Considerations: Version 3, March 22, 2023, 1, https://www.nerc.com/comm/RSTC_Reliability_Guidelines/Reliability%20Guideline%20-%20Gas%20and%20Electric%20Operational%20Coord%20Considerations.pdf

254 Ibid

255 NERC, Natural Gas, 3

fuel versus natural gas. However, the Department of Homeland Security's Transportation Security Administration (TSA) issued new cybersecurity incident reporting requirements for TSA-designated critical pipelines that transport hazardous liquids and natural gas. In particular, a May 2021 Security Directive by TSA required that critical pipeline owners and operators report confirmed and potential cybersecurity incidents to the DHS Cybersecurity and Infrastructure Security Agency (CISA).[256] Subsequent revisions to that Directive have provided additional requirements on reporting and related issues.[257]

That Directive and its revisions, however, do not address the issue of whether and how TSA or CISA would provide timely incident notification to the electric utilities and their subsector partners (including RCs) of impending natural gas supply interruptions that could affect blackstart. Such notification will be essential as utilities assess the availability of blackstart infrastructure (including generation) and determine whether contingency plans may be needed for restoration to go forward.

The Argonne National Laboratory's NGInsight tool is designed to improve situational awareness between natural gas pipeline and BES operators; DOE might fund its further development and deployment to facilitate incident notification and emergency coordination.[258] Another means by which utilities might receive such information is via an Electricity Subsector Coordinating Council (ESCC) call and other subsector emergency communications operations (including under the NERC Alert system[259]) that would already be

256 Department of Homeland Security (DHS), DHS Announces New Cybersecurity Requirements for Critical Pipeline Owners and Operators, May 27, 2021, https://www.dhs.gov/news/2021/05/27/dhs-announces-new-cybersecurity-requirements-critical-pipeline-owners-and-operators. A subsequent Security Directive in July 2021 required these critical pipelines to adopt additional cybersecurity measures. https://www.tsa.gov/news/press/releases/2021/07/20/dhs-announces-new-cybersecurity-requirements-critical-pipeline

257 Department of Homeland Security (DHS), Memorandum: Renewal With Revisions to the Security Directive Pipeline 2021-02 Series, Cybersecurity Mitigation Actions, Contingency Planning, and Testing, July 26, 2023, https://www.tsa.gov/sites/default/files/tsa-sd-pipeline-2021-02d-w-memo_07_27_2023.pdf

258 North American Energy Standards Board, Gas Electric Harmonization Forum Report July 28, 2023, https://www.naesb.org/pdf4/geh_final_report_072823.pdf

259 North American Electric Reliability Corporation (NERC), About Alerts, https://www.nerc.com/pa/rrm/bpsa/Pages/About-Alerts.aspx

underway in response to the original, wide-area power outage that necessitates blackstart. ESCC calls could include the ONG-Subsector Coordinating Council leadership, the American Gas Association (AGA), and the Interstate Natural Gas Association of America (INGAA) as well as other ONG industry groups as appropriate to the event. As an initial step, the ESCC and its partners explore how such existing call mechanisms could be used to promptly share notifications of natural gas pipeline disruptions that would affect restoration. As DOE continues to strengthen Emergency Support Function (ESF) #12-Energy, the ONG and electricity subsectors might also explore whether and how ESF calls might supplement other mechanisms to establish shared situational awareness as the emergency unfolds.

Emergency communications between utilities and natural gas pipeline operators will also need to extend long beyond the initial notification of an event. Indeed, real time communications between them will be essential throughout the restoration process. TOPs and pipeline operators cannot act and respond independently. The control centers for pipelines, generators, and TOPs have to appreciate how their decisions impact the operations of the others and coordinate their actions in real time across their systems.

TOPs and pipeline operators cannot act and respond independently. The control centers for pipelines, generators, and TOPs have to appreciate how their decisions impact the operations of the others and coordinate their actions in real time across their systems.

Data must also flow in both directions. TOPs need awareness from pipeline companies of the operational measures they are taking. For example, to the extent that a TOP needs to direct generator operations for power restoration, it may not be possible if the supplying pipeline has restoration delivery planned for elsewhere, or if a pipeline's scheduling cycle requirements are out of sync with restoration support requirements. At the same time, it will be essential to include pipeline operators in emergency communications and shared situational awareness, and in changes in TOP and generator operations so all such partners can account for those changes in their own incident management decision-making.

In addition, if a region has pipelines that have not been disrupted by an attack, and can support blackstart and grid stabilization, the operation of those surviving pipelines could be impacted by TOP directives to generators and should therefore be included in emergency communications throughout the event. It will also be necessary to prioritize the delivery (or rapid restoration) of power to the compressor stations and associated infrastructure of those unimpacted pipelines.

Furthermore, these communications protocols and processes will need to account for whether utilities conduct centralized generation dispatch (and can therefore coordinate fairly easily with generators and pipeline operators across the affected region), or whether multiple entities and independently-owned generators will be involved in dispatch for the region (requiring additional real-time information sharing and coordination mechanisms and, potentially, raising antitrust issues that should be addressed before such events occur).

In addition to receiving timely incident reports, additional data might help TOPs establish situational awareness prior to and during initial restoration operations and adjust their blackstart plans accordingly. Even after a catastrophic cyberattack on transmission pipelines, the linepack of gas remaining in the pipelines constitutes de-facto storage, and availability of that gas to serve power generators and other customers will gradually decline rather than immediately halt.[260] Access to such remaining linepack early in the blackstart process could enable the use of innovative contingency plans. However, given the large volumes of natural gas necessary to start up and operate many cranking path generators, further research and discussions with interstate natural gas pipeline operators will be necessary to determine whether and how such emergency linepack utilization might be possible.

Sharing additional data on the status and deliverability of stored natural gas to specific generators could also be helpful for blackstart contingency planning. Identifying the types of data and emergency support that would benefit restoration, above and beyond the timely notification that an incident has occurred, should become a focus of future natural gas-electric discussions.

Measures will also be needed to better coordinate the allocation of available gas supplies. If attackers disrupt interstate gas pipelines, a huge gap will emerge

260 For a concise overview of linepack and related natural gas system characteristics, see AGA, How Does the Natural Gas System Work, https://www.aga.org/natural-gas/delivery/how-does-the-natural-gas-delivery-system-work-/

between available gas supplies and demand for them. Recognizing the contractual delivery and supply obligations in place, the electricity subsector should intensify discussions with the natural gas industry and government partners on criteria to prioritize the allocation of limited natural gas supplies, especially between generators and local natural gas distribution companies (LDCs).

Recognizing the contractual delivery and supply obligations in place, the electricity subsector should intensify discussions with the natural gas industry and government partners on criteria to prioritize the allocation of limited natural gas supplies, especially between generators and local natural gas distribution companies (LDCs).

Bilateral communications between interdependent electric and natural gas operators can provide a more resilient means of coordination during an emergency as compared with reliance on central communications from FEMA, TSA, or CISA. One framework for this coordination is the "Continuity of the Economy" (COTE) legislative framework for cyber-physical systems that is under consideration by Congress.[261] This calls for electric and natural gas utilities to conduct network interdependency analysis and to identify all outside organizations upon which they depend for emergency operations, with prioritization of those organizations and facilities that are particularly critical for resilient, reliable, and adequate operations. Bilateral emergency communication and recovery plans can be developed by these partner operators in the absence of centrally coordinated action, or as a backup contingency for when that central coordination fails during the chaos of a blackout or through government inaction.

261 Ruddell, B. L. (2023) Rationale and Process for Continuity of the Economy. (Report No. IHS/CR-2023-1021). The Sam Houston State University Institute for Homeland Security; https://ihsonline.org/Portals/0/Tech%20Papers/2023_Papers/Ruddell_2023_Continuity_of_the_Economy.pdf

IV | SECONDARY FUELS AS A SPECIAL FOCUS FOR ACTION

The ability of dual-fuel generators to run on diesel instead of natural gas bolsters blackstart system resilience – but only if fuel distribution companies can resupply them. As noted above, in a wide-area backout, fuel providers will face overwhelming requests for diesel deliveries in a severely disrupted environment. Moreover, even in normal conditions, many generators will only be able to receive deliveries sufficient for several hours of operations per day. A wide-area blackout will greatly impede delivery operations. Utilities and their ONG and government partners should develop options to meet these challenges, including emergency waivers of transportation and emissions regulations to facilitate secondary fuel operations for blackstart.

The ability of dual-fuel generators to run on diesel instead of natural gas bolsters blackstart system resilience – but only if fuel distribution companies can resupply them.

However, these generators typically have only limited supplies of fuel stored on-site. Resupplying them with secondary fuel will be critical until natural gas supplies are restored. A number of factors could impede the resupply of secondary fuel to these generators. First, in the event of a wide area outage, enormous demand will emerge for fuel not only from blackstart generators but also from the hospitals, water and wastewater systems, cell towers and other communications system facilities, nursing homes, and a vast array of other facilities that have emergency power generators and provide critical services. Few if any fuel distribution companies have scaled their truck fleets and other infrastructure to meet such a tidal wave of requirements. Second, as with all facets of restoration in a Black Sky environment, resupply operations will need to go forward in a severely disrupted environment. For example, the power outage could produce large-scale failures of traffic signals, and bad weather could exacerbate any such traffic problems.

TOPs should discuss these resupply issues with their generation operators and fuel distribution companies to better assess the risks they face, and to develop near- and long-term plans to mitigate these risks. Establishing baseline requirements for resupply of generators will be a key starting point for such discussions. One participant estimated that resupplying a single generator could require up to one fully loaded truck per hour. Moreover, as with allocation of limited natural gas resources in a wartime environment, no criteria yet exist for how to prioritize secondary fuel distribution, and deal with the competition for limited diesel supplies between multiple types of critical customers.

One participant estimated that resupplying a single generator could require up to one fully loaded truck per hour.

DOE and the ONG SCC will be valuable partners to help gather data on secondary fuel distribution capabilities and challenges in a disrupted environment, and for developing allocation criteria. Including DHS will be helpful as well, given the impact of emergency power disruptions on systemically important critical infrastructure (SICI) and national critical functions.[262] Broader federal and state participation in such prioritization effects may be helpful as well, including to address potential ant-trust issues that these initiatives may create as well as understanding state-level priorities.

Discussions should also go forward with government partners to pre-plan for the use of waivers to support generator resupply operations. In particular, preplanning for two types of waivers might be helpful: 1) waivers of environmental regulations concerning the duration and scale for burning secondary fuels; and 2) waiver of local, state, and federal transportation regulations and inspection requirements that might otherwise slow the delivery of such fuels. Many government agencies already have an extensive array of playbooks and waiver authorities for use in other types of emergencies. Conducting additional utility-government exercises on the use of these authorities for restoration support could be helpful in identifying opportunities to meet other challenges, including the credentialing of trucks/drivers to be able to access fuel terminals even if they do not normally perform such functions.

> The purpose of this Handbook is to provide an easy-to-use, comprehensive resource for cost effective EMP protection measures as part of an overall system planning process.

262 For a definition of systemically important critical infrastructure (SICI) and the rational for establishing this category of infrastructure, see the Cyberspace Solarium Commission, Final Report, March 2020, p. 97, https://drive.google.com/file/d/1ryMCIL_dZ30QyjFqFkkf10MxIXJGT4yv/view

V | REGULATORY AND COST RECOVERY INITIATIVES IN A DECARBONIZING GRID

Measures to create blackstart exceptions to state and federal zero-carbon policies will be necessary but not sufficient to help ensure that gas-fired generators can continue to support restoration. As decarbonization begins to downsize the broader gas generation fleet, economic pressures are likely to grow on gas system owners and operators to downsize their supporting infrastructure as well, especially as electrification drives down the demand from other sources, including manufacturing and home heating. We need to ensure the adequacy of the full

As decarbonization begins to downsize the broader gas generation fleet, economic pressures are likely to grow on gas system owners and operators to downsize their supporting infrastructure as well.

range of facilities and operations necessary to fuel gas-fired blackstart assets, unless and until such time as GFM IBRs can fully and effectively provide the power necessary for restoration.

An important opportunity has emerged to help meet such requirements. As the U.S. Environmental Protection Agency (EPA) was formulating its May 2023 proposal to sharply reduce carbon emissions by coal and natural gas-fired generation, the EPA and Department of Energy (DOE) agreed to cooperate "on electric sector resource adequacy and operational reliability."[263] The agreement also provides for regular outreach and consultation with FERC, NERC, state PUCs and other resilience stakeholders. As part of their collaborative efforts, these partners should join together with the Oil and Natural Gas (ONG) subsector to explore measures to retain the infrastructure necessary to fuel blackstart operations.

However, discussions between the stakeholders will also need to address more immediate challenges to grid reliability. In comments jointly filed by RTOs and ISOs across the United States, these organizations warned the proposed EPA rule "could place the reliability of the electric grid in jeopardy." In particular:

The Joint ISOs/RTOs are concerned that the substance of the Proposed Rule as presently configured, as well as its timing, have the potential to materially and adversely impact electric reliability. Moreover, the Proposed Rule, when combined with other EPA rules and other policy actions, could well exacerbate the disturbing trend and growing risk wherein the pace of retirements of generation with attributes needed to ensure grid reliability is rapidly exceeding the commercialization of new resources capable of providing those reliability attributes.[264]

263 U.S. Department of Energy and U.S. Environmental Protection Agency, Joint Memorandum on Interagency Communication and Consultation on Electric Reliability, March 9, 2023, 1-2,https://www.epa.gov/system/files/documents/2023-03/DOE-EPA%20Electric%20Reliability%20MOU.pdf

264 Electric Reliability Council of Texas, Inc.; Midcontinent Independent System Operator, Inc.; PJM Interconnection, LLC; and Southwest Power Pool, Inc., Joint Comments on the EPA's Proposed New Source Performance Standards for Greenhouse Gas Emissions, August 8, 2023, https://pjm.com/-/media/documents/other-fed-state/20230808-comments-of-joint-isos-rtos-docket-epa-hq-oar-2023-0072.ashx

The RTO/ISO comments do not mention the risks that the retirement of gas-fired generators could pose to blackstart restoration. However, given the critical role that these assets play for next start and subsequent phases of blackstart restoration, and the need to maintain the supporting infrastructure to keep them effectively operating, the EPA's proposed retirements are concerning indeed. The RTOs and ISOs note that they "are already seeing retirements of generators that are concerning as they appear to be driven by a reluctance of investors to make the commitments needed to keep these capital-intensive resources operating." Moreover, "as the penetration of renewable resources continues to increase, the grid will need to rely even more on generation capable of providing critical reliability attributes."[265] Equivalent attributes will be essential to help maintain the stability of blackstart power islands as they grow and are synchronized with their neighbors.

The RTOs and ISOs note that they "are already seeing retirements of generators that are concerning as they appear to be driven by a reluctance of investors to make the commitments needed to keep these capital-intensive resources operating.

To mitigate the potential risks that the EPA rule poses to reliability, the RTOs and ISOs propose creating an exception to the fossil retirements that the proposed EPA rule would require - a recommendation that could also be applied to meet blackstart needs as well. These organizations propose that the EPA establish a new "reliability-based sub-category" that would provide a time-limited means for ISOs/RTOs to designate classes of existing units needed to maintain local or region-wide reliability until alternatives (which might be new transmission or new generation or storage resources) are available to address the identified reliability need. The ISO/RTO would identify these units or locations to EPA and a unit's placement in this sub-category would allow the noncompliant units to continue to operate beyond the date of compliance with the rule until alternative solutions can be placed into service.[266]

265 Joint Comments, 2

266 Under this proposal, each ISO/RTO would provide a public explanation of the methodology it would use to determine which units, or classes of units, qualify for inclusion in this subcategory and the

That same sub-category could be structured to include blackstart units, next generation assets, and other fossil resources that might otherwise be shut down under the EPA's proposed rule. Stringent measures would be needed to protect the data associated with the blackstart assets slated to be kept in service under this provision. In addition, as GFM resources (and especially BESS) become increasingly capable of supporting blackstart restoration, specialized technical and operational assessments will be needed to identify opportunities to phase out gas units over time to meet decarbonization goals. But the overall strategy is sound: create exceptions as needed for blackstart because blackstart is so exceptionally important in an era of Black Sky hazards.

process for identification of such units. The ISO/RTO would then conduct a unit or location-specific reliability analysis for each of these units. The analysis would establish the defined period past the initial retirement date that the unit is needed to maintain grid reliability while measures are implemented to address reliability issues caused by the affected unit's retirement. Joint Comments, 3, 10

VI | MANDATORY RELIABILITY AND CRITICAL INFRASTRUCTURE PROTECTION STANDARDS FOR NATURAL GAS SYSTEMS

Winter Storm Elliott exemplifies the need for mandatory reliability standards for natural gas systems. The storm brought below-freezing temperatures to major portions of the eastern United States for several days in December 2022. FERC has reported that the natural gas system in New York City nearly failed and left a million customers without heat on Christmas Eve. Had Consolidated Edison Inc. — New York State's largest utility — not taken emergency action, gas heating could have been shut off for months in "all or portions" of its territory, according to staff at the Federal Energy Regulatory Commission. The utility provides gas service to 1.1 million customers in Manhattan, the Bronx and parts of Queens and Westchester County.[267]

The impact of such an extended gas outage in New York City and surrounding areas on residential heating and other gas customers would have been catastrophic. However, from the perspective of grid reliability, the risks of such gas supply interruptions or outages are extraordinarily concerning as well. Outages at 63 natural gas-fueled power plant units resulted from natural gas deliveries being cut off, FERC's analysis found.[268] The longer duration disruptions that Elliot nearly created would have had profound effects of power generation in the region.

NERC and FERC officials noted that such disruptions are recurrent events. Winter Storm Uri, for example, left millions of customers in Texas without power and more than 200 people dead in February 2021. Jim Robb, CEO of NERC, noted: "Five events in 11 years, three events in the last five — that's too often. These are not infrequent events."[269]

The obvious solution: move from voluntary compliance with guidelines to mandatory reliability standards, equivalent to those established for the reliability of the BES. FERC Chairman Willie Phillips has emphasized that "Someone – it doesn't have to be FERC – must have authority to establish and enforce natural gas reliability standards." He also noted that "some recommendations from the 2021 Uri report are still not implemented."[270] Enforcement of such standards will also be essential to help ensure the resilient flow of fuel for power generation in severe weather events.

Mandatory standards will also be needed for blackstart generators vital for all phases of the restoration process. The FERC/NERC report on Elliot recommended that "NERC and its Regional Entities, with Commission staff engagement, should initiate a study modeled after the 2021 Report Rec. 26/ ERCOT blackstart unit availability study, to assess readiness of other blackstart

267 Miranda Wilson, "FERC Outlines 'Catastrophic' Gas, Grid Threats," EE News, September 22, 2023, https://www.eenews.net/articles/ferc-outlines-catastrophic-gas-grid-threats/

268 Ibid

269 Ibid

270 FERC, Elliott Report: Complete Electricity Standards, Implement Gas Reliability Rules, https://www.ferc.gov/news-events/news/elliott-report-complete-electricity-standards-implement-gas-reliability-rules

units to operate during cold weather conditions, including recommended actions for improvement, (if necessary)."[271]

However, while reliability standards for weather events are necessary, natural gas system operators and their Federal partners should also follow the NERC/FERC lead in establishing Critical Infrastructure Protection mandates for such systems.

Recommendation: The Oil and Natural Gas Subsector should begin now to develop mandatory infrastructure protection standards for natural gas infrastructure before a Black Sky event makes them inevitable – and too late.

The Oil and Natural Gas Subsector should begin now to develop mandatory infrastructure protection standards for natural gas infrastructure before a Black Sky event makes them inevitable – and too late.

In the case of BES reliability and CIP standards, it took a catastrophic outage before the electricity subsector, driven in part by congressional legislation, transitioned from voluntary to mandatory, enforceable requirements. Gas system operators now have the opportunity to lead the development of such standards, supported by specific Intelligence Community findings on attack vectors and mitigation options.

271 FERC/NERC, December 2022 Winter Storm Elliott Grid Operations: Key Findings and Recommendations, Page 17; https://ferc.gov/sites/default/files/2023-09/Presentation%20-%20Elliott_F-R_Open_Mtg_%289-21-23%29.pdf

6

CHAPTER SIX

EMPLOYING INVERTER-BASED RESOURCES FOR BLACKSTART RESTORATION

INTRODUCTION

The FERC-approved tariff proposal by PJM Interconnection, LLC on Black Start Service Revisions to Enhance Fuel Assurance offers an especially detailed and well-documented assessment of options to use IBRs as blackstart units. In particular, the proposal sets criteria for selecting IBRs and other types of generation (including gas-fired generators) as "Fuel Assured Black Start Units." The additional resilience provided by these units could be especially valuable against Black Sky hazards. Furthermore, though not explicitly addressed by PJM, the deployment of numerous, widely dispersed IBRs as black start units could enable TOPs to develop restoration options for improved survivability against coordinated attacks.

The deployment of numerous, widely dispersed IBRs as black start units could enable TOPs to develop restoration options for improved survivability against coordinated attacks.

Under PJM's proposal, fuel assured units would need to offer a minimum runtime of 16 hours, regardless of the type of generation employed.[272] That extended runtime will enable blackstart units to conduct multiple restarts of the restoration process, and thereby mitigate the risks of island instability and collapse. Restart capabilities will be all the more crucial if adversaries conduct sustained Black Sky attacks to disrupt restoration operations.

PJM's analysis highlights an additional benefit of employing fuel assured black start units. TOPs have redundancy built into their restoration plans, so that if the primary cranking path cannot function due to the failure of its blackstart unit or for other reasons, operators can employ alternative cranking paths that rely on different units. Again, given the likelihood that Black Sky events will damage restoration infrastructure, this redundancy offers significant value.

However, falling back on alternative cranking paths entails disadvantages. PJM found that "the required use of alternate cranking paths resulted in increased restoration complexity, and additional restoration time." The primary path is typically the path with the shorter electrical distance from the black start unit to what PJM terms the next "critical load" generator, which helps establish the initial power island. In contrast, alternate paths "typically require additional generation or load to control voltages – additional steps, which increase restoration time."[273]

That time differential is significant. PJM found that the average incremental time increase per cranking path ranged from approximately five hours to 11 hours across the zones that PJM examined. Moreover, about 30 percent of those zones showed incremental increases of between 15 and 40 hours.[274] A delay of up to 40 hours could have disastrous effects on restoration of service to hospitals, water systems, military bases, and other priority loads. The availability of assured fuel blackstart units could help ensure the operational effectiveness of primary cranking paths and avoid such delays.

272 PJM, Black Start Service Revisions, 17-118

273 PJM, Black Start Service Revisions, 11-12

274 PJM, Black Start Service Revisions, 13

Gas Disruptions, IBRs, and Broader Implications for Blackstart Resilience

The primary focus of PJM's proposal is to assess and mitigate the risk of disruptions in the flow of fuel to natural gas-fired black start units. The proposal notes that "fuel supply availability could be a very real concern during the type of severe and widespread disruption to the BES that triggers a need for black start generation," and that units deprived of fuel would be incapable of initiating the restoration process.[275]

This assessment understates the severity of the threat to gas supplies and the consequences for blackstart. The 2023 Annual Threat Assessment of the U.S. Intelligence Community assesses that "If Beijing feared that a major conflict with the United States were imminent, it almost certainly would consider undertaking aggressive cyber operations against U.S. homeland critical infrastructure," aimed in part at "inducing societal panic. Gas infrastructure would offer a prime target for attack. The 2023 assessment emphasizes that "China almost certainly is capable of launching cyberattacks that could disrupt critical infrastructure services within the United States, including against oil and gas pipelines, and rail systems."[276] Such disruptions would not only jeopardize the reliability of gas-fired black start units but also of gas generators on which TOPs rely for next start and subsequent phases of restoration.

The Transportation Security Administration has imposed increasingly stringent regulations to improve the security of natural gas pipeline systems.[277] Over and above these requirements, many system owners and operators are implementing voluntary measures to further strengthen their resilience, including a partnership with the electricity subsector to participate in the ESCC's

275 PJM, Black Start Service Revisions, 9

276 Office of the Director of National Intelligence, Annual Threat Assessment of the U.S. Intelligence Community, February 6, 2023, https://www.dni.gov/files/ODNI/documents/assessments/ATA-2023-Unclassified-Report.pdf; Cybersecurity and Infrastructure Security Agency (CISA), Chinese Gas Pipeline Intrusion Campaign, 2011 to 2013, July 21, 2021, https://www.cisa.gov/news-events/cybersecurity-advisories/aa21-201a

277 Transportation Security Administration, TSA updates, renews cybersecurity requirements for pipeline owners, operators, July 26, 2023, https://www.tsa.gov/news/press/releases/2023/07/26/tsa-updates-renews-cybersecurity-requirements-pipeline-owners

cyber mutual assistance program.[278] Section F examines these collaborative efforts and provides additional options for gas-electric coordination to facilitate blackstart in a disrupted environment.

Other black sky hazards pose equivalent threats to the flow of natural gas. In previous studies, PJM has noted the importance of physical attacks against pipeline infrastructure as scenarios to guide resilience planning.[279] Those threats continue to intensify. The Transportation Security Administration's (TSA) 2020 Biennial National Strategy for Transportation Security identifies pipelines as vulnerable to cyberattacks and physically "vulnerable to terrorist attacks largely due to their stationary nature, the volatility of transported products, and [their] dispersed nature."[280]

The Transportation Security Administration's (TSA) 2020 Biennial National Strategy for Transportation Security identifies pipelines as vulnerable to cyberattacks and physically "vulnerable to terrorist attacks largely due to their stationary nature, the volatility of transported products, and [their] dispersed nature.

The PJM tariff offers detailed criteria for gas-fired generators to qualify as Fuel Assured Back Start Units. Where on-site fuel storage is not feasible for such resources, a gas unit will qualify as fuel assured if it has independent connections to two interstate pipelines, or if it has a direct connection to a gas gathering system. Direct connection to a natural gas gathering system "ahead of any interstate pipeline" also provides the higher level of fuel assurance that the Black Start Unit's responsibilities demand, by avoiding or reducing the risk of pipeline-ordered curtailments.[281] Equivalent fuel assurance measures could also help

278 American Gas Association (AGA), AGA Cybersecurity Initiatives, https://www.aga.org/natural-gas/security/cybersecurity/aga-cybersecurity-initiatives/

279 PJM, Fuel Security Analysis: Summary of Discussions with the U.S. Department of Energy and Requested Analyses, 2018, https://www.pjm.com/-/media/library/reports-notices/fuel-security/fuel-security-summary-of-discussions-with-federal-agencies-and-requested-analyses.ashx?la=en

280 Transportation Security Administration (TSA) 2020, Biennial National Strategy for Transportation Security May 29, 2020, https://www.dhs.gov/sites/default/files/publications/2020-06-12_2020-biennial-national-strategy-transportation-security-report.pdf

281 PJM, Black Start Service Revisions, 19-20

strengthen the resilience of gas-fired generation for next start and follow-on restoration operations.

In one respect, IBRs offer an inherent advantage over gas-fired generators: the supply of "fuel" for solar and wind power cannot be intentionally cut off. However, given the intermittency of the electricity provided by solar photovoltaic panels and wind turbines, their ability to serve as blackstart units (and as reliable sources of power for subsequent restoration phases) is very much in question. Dominion Energy highlights these variability problems in its 2023 Integrated Resource Plan. Dominion states that "at this point in time, solar generation would not be used for black start system restoration due to the impacts intermittent generation would have on grid stability during blackstart system restoration. Until there is sufficient energy storage to generate electricity at night and to mitigate the impacts of intermittent generation, solar generation will provide little to no value for blackstart purposes." But Dominion also emphasizes that "the Company stands ready to meet the challenges associated with solar generation with continued study, technological advancement, and innovation" in the years to come.[282]

Emerging technologies may soon improve the availability of solar and wind-provided power. Given the pace of technological change, the best approach is to specify the functional requirements that IBRs and alternative sources of power must meet to be used as blackstart resources, and then "let the chips fall" in an objective, equitable way.

As will be discussed below, emerging technologies – especially those associated with long-duration energy storage systems and advanced inverters to control the electricity they discharge – may soon improve the availability of solar and wind-provided power. Given the pace of technological change, the best approach is to specify the functional requirements that IBRs and alternative sources of power must meet to be used as blackstart resources, and then "let the chips fall" in an objective, equitable way.

282 Dominion Energy, 2023 Integrated Resource Plan, May 1, 2023 84, 2023 Integrated Resource Plan Virginia Electric and Power Company (azureedge.net)

PJM has proposed just such a strategy. In its filing to FERC, PJM specified the criteria that intermittent and hybrid resources must meet to serve as Fuel Assured Black Start Units. Such units must be able to provide the required power capacity that, based on the unit's historic operations, is 90 percent certain to be able to sustain operations for a total of 16 hours. PJM's filing also documents why these criteria are functionally equivalent to the standards it applies for gas generation or other fuel assured options and provides a fair basis for selecting blackstart units among these alternatives.[283]

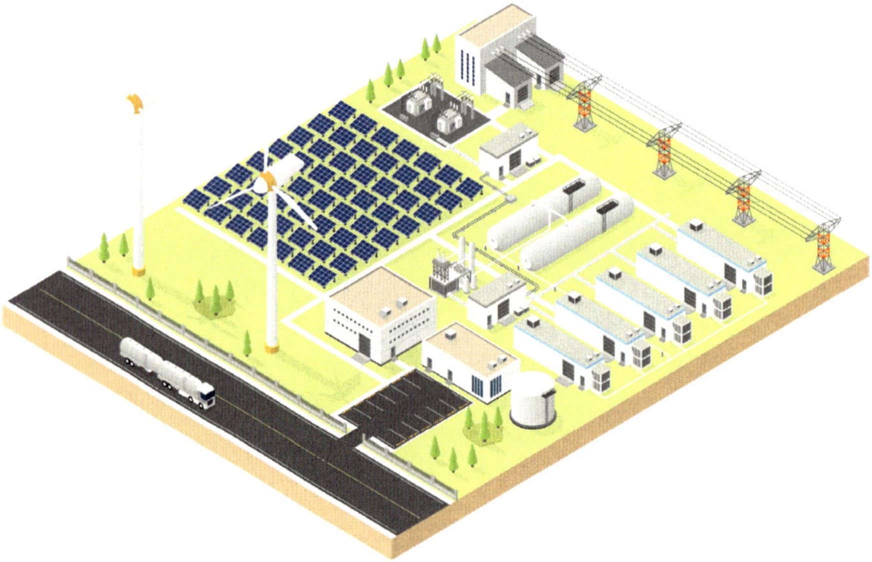

283 PJM, Black Start Service Revisions, 20-22

I | IBRS FOR BLACKSTART – POTENTIAL RESILIENCE BENEFITS

Hybrid plants are the most likely IBRs to meet PJM's criteria, or equivalent ones that other RTOs, ISOs, and transmission operators are considering. New technologies are extending the storage duration and capacity of utility-scale BESS, and the pairing of multiple batteries with solar and wind recharging capabilities can provide for extended delivery of power.[284] Demonstration projects have been underway for years to assess the value of hybrid facilities and their advanced batteries for blackstart. In 2017 For example, the Imperial Irrigation District (IID) utility in Southern California successfully used a 33MW/20MWh lithium-ion battery energy storage system (BESS) to kick-start

284 PJM also proposes that run-of-river generation be eligible to serve as Fuel Assured Blackstart Units, other forms of generation might be considered as well, including pumped hydro and advanced energy storage systems that meet PJM's criteria.

an 44MW combined cycle natural gas turbine, located at El Centro Generating Station in Imperial Valley, California.[285] With continuing improvements in BESS capabilities, many additional demonstration projects and operational deployments to support restoration are now going forward worldwide.

Significant challenges remain to replacing gas-fired black start units with hybrid plants. If the batteries in those plants are to serve as blackstart units, their stored power must be reserved for that function. Batteries will be useless for energizing cranking paths if their power has already been discharged for normal electricity dispatch and market operations. Specialized financial incentives and associated contract terms of service will be essential to help generator owners (GOs) recoup their investments in hybrid plant blackstart units.

Moreover, as with all blackstart system components, hybrid plants and other IBRs must be able to survive the initial Black Sky event that necessitates restoration and continue to function in the face of the follow-on attacks that adversaries may conduct. Assuming that these requirements can be met, however, hybrid plants that offer 90% confidence of providing 16 hours of power (and are therefore capable of multiple restarts) will provide an increasingly useful option for initiating blackstart restoration.

Batteries will be useless for energizing cranking paths if their power has already been discharged for normal electricity dispatch and market operations.

IBRs offer other potential benefits for Black Sky resilience as well. The most obvious is diversity of generation resources: if blackstart systems employ multiple types of assets, they will be less susceptible to common mode failures, and will complicate adversary attack planning and execution. Adding diversity to systems that are nearly or totally dependent on gas-fired blackstart units and next start resources offers an especially important target of opportunity, especially given China's focus on attacking natural gas infrastructure.

Diversification also provides a hedge against coordinated, multi-pipeline

285 Andy Colthorp, California battery's black start capability hailed as 'major accomplishment in the energy industry,' Energy Storage News, May 17, 2017, https://www.energy-storage.news/california-batterys-black-start-capability-hailed-as-major-accomplishment-in-the-energy-industry/

attacks on that infrastructure. To qualify as a Fuel Assured Blackstart Unit, gas-fired units that cannot employ stored on-site fuel would need to connect to more than one interstate gas pipeline. That requirement reflects an implied design basis threat: i.e., adversaries are assumed to be able to disrupt one such pipeline, so being served by two or more is necessary to meet PJM's fuel assurance criteria.

However, nothing in the 2023 ODNI Threat Assessment suggests that China will limit itself to attacking two pipelines in future confrontations. On the contrary: the Assessment states if China feared that a major conflict with the United States was imminent, it would strike U.S. homeland infrastructure to coerce U.S. decision-making and incite societal panic.[286] Comprehensive attacks against the interstate natural gas pipeline system would be best suited to achieve these goals, though we should also be prepared for initial "demonstration" strikes that illustrate the vulnerability of that system and encourage the President to settle the conflict before attacks escalate.

To qualify as a Fuel Assured Blackstart Unit, gas-fired units that cannot employ stored on-site fuel would need to connect to more than one interstate gas pipeline.

In addition to offering a hedge against multi-pipeline attacks, employing IBRs as blackstart units could also increase the redundancy of blackstart cranking paths. While each blackstart unit can only crank one Critical Load at a time, it is possible for multiple units to energize separate cranking paths and power islands in parallel.[287] The ability to do so would help mitigate the risk that the initial Black Sky event that requires blackstart restoration will damage the generation or transmission infrastructure of one or more cranking paths (including, potentially, both the primary and secondary "alternate" paths). Redundant paths can also strengthen system resilience if adversaries specifically target restoration infrastructure with either cyber or combined cyber-physical attacks. Indeed, given the importance of blackstart to national security, it is prudent to assume that blackstart assets will be our adversaries' high priorities for destruction.

286 ODNI, 2023 Threat Assessment, 10.

287 PJM, Black Start Service Revisions, 20-22

II | MULTI-ISLAND PLANNING AND OPERATIONS

The wide dispersal of IBRs across the U.S. and partner nations will create new opportunities for TOPs and their partners to maneuver around disrupted assets. Given the value of hybrid black start units for conducting multiple restarts, their increasing deployment could offer particular benefits for enabling the use of multiple cranking paths. At the end of 2022, there were 374 hybrid plants (>1 MW) operating across the United States, totaling nearly 41 GW of generating capacity and 5.4 GW/15.2 GWh of energy storage (over a 70% increase from 2021). Moreover, proposals for new construction of hybrid plants are rapidly growing. At the close of 2022, there were 51% more hybrid plants—representing 59% more generating capacity—in interconnection queues across the United States than there were at the end of 2021. The average storage duration for these plants was only 3.1 hours.[288] However, as noted above, new battery technologies and other advances in storage will enable new plants to achieve much longer duration.

Moreover, depending on the topography of transmission and distribution infrastructure, and connectivity they provide to critical and priority loads, the use of widely distributed IBR units could help TOPs establish multiple power islands at the same time in ways that accelerate overall restoration timelines. PJM's analysis found that using backup cranking paths would take up to 40 hours longer than primary paths to energize the next critical load, with added restoration complexity. Yet, with the widespread construction of hybrid plants and their use as fuel-assured units, it may become possible to plan for the simultaneous establishment of multiple power islands, and the more rapid restoration of service to critical/primary loads and next start generators than would be possible by using a single cranking path. Multi-island plans could also:

With the widespread construction of hybrid plants and their use as fuel-assured units, it may become possible to plan for the simultaneous establishment of multiple power islands, and the more rapid restoration of service to critical/primary loads and next start generators than would be possible by using a single cranking path.

- mitigate the danger that the next start units that are supposed to be energized by the primary cranking paths will be disrupted, and thereby impede restoration even if the cranking path itself is operational, and
- help ensure that restoration goes forward if the power island created by the primary cranking path collapses due to frequency or voltage instabilities, necessitating a time-consuming restart of the process to energize that path.
- Mitigate the severity of the original grid outage and reduce the extent of the blackout - thus boosting resilience - by placing island boundaries strategically so that cascading power grid failures are naturally

288 Larence Berkeley National Laboratory, Hybrid Power Plants, https://emp.lbl.gov/hybrid

interrupted as they propagate from one grid island to another. Careful placement of grid-forming microgrids at the borders of islands and at critical nodes that participate in many cranking paths and many cascading failure paths is a practical means for a plan to achieve this objective.

Simultaneously establishing multiple power islands would create new burdens for grid operators. TOPs and their partners would have to replicate the same tasks needed for a single island at a variety of locations, each with their own specific topographies and challenges, including for 1) segmenting these islands from surrounding high voltage transmission and distribution-level infrastructure, 2) sequencing their pickup of load and generation, and 3) maintaining their stability and synchronizing them with a larger number of neighboring islands.

Despite these difficulties, some TOPs with large service areas already have plans to establish multiple islands or are exploring options to do so. Utilizing increasingly dispersed IBRs as blackstart units (along with other fuel-assured resources) could facilitate their efforts. At present, however, many GOs face pressures in the opposite direction: not to expand the availability of blackstart resources, but rather to cut back on them due to financial and regulatory concerns.

III | BLACKSTART COST RECOVERY, REGULATIONS, AND SECURITY IN A CHANGING RESOURCE MIX

Local, state, and federal initiatives to reduce carbon emissions (and in some cases mandate a zero-carbon grid) provide only one of the downward pressures on the availability of gas-fired generators for blackstart restoration. Ensuring adequate compensation and cost recovery for generation owners in organized markets will also be crucial. The costs of complying with cyber and physical security regulations, and the risk that generation owners will need to pay penalties for non-compliance, creates further availability issues. Yet, given the intensifying threats to the grid and danger that adversaries prioritize the disruption of blackstart resources and operations, we also need to examine whether regulations for such resources are sufficiently stringent and comprehensive.

Striking the appropriate balance between these conflicting pressures will be difficult for gas-fired resources. For IBRs, such issues constitute terra incognito. Little analysis has been performed as to whether and how NERC physical and cyber security standards ought to be applied to hybrid plants that serve as blackstart units. The consideration of such standards (and of financial incentives) for next start assets is equally under-studied, even though retirements of gas generators are accelerating and projects to replace that lost capacity often face strong political and regulatory headwinds.[289] The analysis that follows addresses these issues for both blackstart units and next start generation, and suggests options to help maintain their adequacy and resilience against Black Sky hazards.

B. Blackstart Resources: Security Versus Availability

NERC's Supplemental Material for CIP-002-6 - Cyber Security - BES Cyber System Categorization highlights the competing pressures that regulators and their partners face in ensuring both the adequacy and survivability of blackstart resources. To comply with that CIP standard, Balancing Authorities, Generator Operators, Transmission Operators, and other functional entities must:

> *Identify and categorize BES Cyber Systems and their associated BES Cyber Assets for the application of cybersecurity requirements commensurate with the adverse impact that loss, compromise, or misuse of those BES Cyber Systems could have on the reliable operation of the BES. Identification and categorization of BES Cyber Systems support appropriate protection against compromises that could lead to misoperation or instability in the BES.*[290]

289 See, for example, Patrick Larsen, Dominion's proposed Chesterfield gas plant draws opposition, VPM News, June 28, 2023, https://www.vpm.org/news/2023-06-28/chesterfield-gas-plant-dominion-virginia-clean-economy-act-climate-change

290 CIP 002-6, 1

These functional entities also include Distribution Providers that own "Each Cranking Path and group of Elements meeting the initial switching requirements from a Blackstart Resource up to and including the first interconnection point of the starting station service of the next generation unit(s) to be started."[291] This provision will be increasingly valuable for cybersecurity as PJM and other BES entities make Distributed Energy Resources eligible to serve as blackstart units.

Blackstart resources and cranking paths are among the systems that require classification as either high impact, medium impact, or low impact systems in terms of the consequences for BES reliability if they were disrupted or misoperated. Draft 1 of CIP-002-5 had proposed to reclassify blackstart resources and cranking paths from low to medium impact. However, after input from NERC's Operating and Planning committees, NERC decided not to do so.[292] The most recent version of the standard, CIP-002-6, assigns blackstart resources and cranking paths a low impact rating.

Increased compliance costs and their impact on resource survivability were key factors in that decision. NERC states that in discussing the proposed reclassification to a medium rating, which would require additional expenses for protection and other measures beyond those necessary for low-rated systems, concerns emerged that "entities owning Blackstart Resources and Cranking Paths might elect to remove those services to avoid higher compliance costs. For example, one Reliability Coordinator reported a 25% reduction of Blackstart Resources in response to previous compliance requirements, and that "there could be more entities that make this choice under Version 5."[293]

NERC also determined that rating these resources as medium in Version 5 would "very significantly increase compliance costs and result in further reduction of a vulnerable pool." As a result of these considerations, the standards

291 Ibid

292 NERC, CIP-002-5 — Cyber Security — BES Cyber Asset and BES Cyber System Categorization, 33-35 (deleting BS and cranking path from Medium) and pages starting at page 38 (adding BS to low) in the redline of Draft 2 vs. Draft 1 of CIP-005

https://www.nerc.com/pa/Stand/Project20086CyberSecurityOrder706Version5CIPStanda/CIP-002-5_redline_from_initial_ballot_(2012-0326).pdf;

https://www.nerc.com/pa/Stand/Project20086CyberSecurityOrder706Version5CIPStanda/CIP-002-5_clean_20111107.pdf

293 CIP-002-6, 35

drafting team "moved from the categorization of restoration assets such as Blackstart Resources and Cranking Paths as medium impact (as was the case in earlier drafts) to categorization of these assets as low impact..." Low impact categorization still requires measures for cyber security awareness, physical access control, and electronics access control. NERC states that this classification therefore "represents a net gain to bulk power system reliability" in comparison with early versions of the standard. [294]

Categorizing blackstart resources as low impact reflects a difficult choice over what is most important for grid reliability: strengthening the security of these assets or ensuring their availability.

Categorizing blackstart resources as low impact reflects a difficult choice over what is most important for grid reliability: strengthening the security of these assets or ensuring their availability. NERC succinctly captures this tradeoff:

> *Weighing the risks to overall BES reliability, the drafting team determined that this recategorization represents the option that would be the least detrimental to restoration function and, thus, overall BES reliability. Removing Blackstart Resources and Cranking Paths from medium impact promotes overall reliability, as the likely alternative is fewer Blackstart Resources supporting timely restoration when needed.*[295]

Striking the right balance between availability and security merits continuous reassessment. Given the vital importance of blackstart restoration to national security, and the consequences that would ensue if cyberattacks caused the loss, compromise, or misuse of blackstart resources and cranking paths, it seems problematic to rate blackstart resources and cranking paths as low impact. Given the likelihood that China or other adversaries will prioritize the disruption of blackstart restoration if a conflict occurs, the benefits or requiring more stringent protection of these assets and operations also appears compelling. Yet,

294 CIP-002-6, 35

295 CIP-002-6, Version 35

NERC's supplemental materials make a critical point: if increased cybersecurity costs reduce the availability of blackstart resources, timely restoration could be jeopardized.

Recommendation: One option to resolve this dilemma is to develop new funding strategies to compensate generator owners for higher levels of security than required under a low impact rating. Just as regulatory carveouts in zero-carbon policies may be needed to assure the availability of gas-fired blackstart resources, we should also explore how to increase the financial incentives for GOs to continue providing blackstart services even if their compliance costs grow. Regulators and their partners should also examine whether such incentives should be developed for next start gas generators, given the downward pressures they will face as well as decarbonization goes forward.

Equivalent measures may be needed to deal with the costs of strengthening the physical security of blackstart assets. NERC's CIP-014-3- Physical Security standard is designed to "identify and protect Transmission stations and Transmission substations, and their associated primary control centers, that if rendered inoperable or damaged as a result of a physical attack could result in widespread instability, uncontrolled separation, or Cascading within an Interconnection."[296] How that standard might apply to blackstart infrastructure is unclear. The control centers necessary for blackstart restoration are almost certainly covered by CIP-014-1. However, CIP-014 does not provide explicit physical security requirements for blackstart units and other assets that such restoration would entail, but which if disrupted would not cause uncontrolled separation or meet the other CIP-014 criteria.

Given the risk that adversaries will combine physical attacks with cyber operations (as posited in GridEx threat scenarios), and target combined attacks to disrupt restoration operations, further consideration should be given to whether and how specialized physical security measures should be applied to blackstart assets either within CIP-014 or through alternative approaches. Dave Roop notes that "power plant owners already have problems in ensuring that their legacy plants comply with CIP-014-1, given the physical layout of those

296 NERC, CIP-014-1 - Physical Security, https://www.nerc.com/pa/Stand/Reliability%20Standards/CIP-014-1.pdf

plants and the many entry points that one could use to penetrate them. Newer plants are better suited for compliance but still tend to have many entry points, complicating lockdown and access control." Measures to remedy these security risks can add significantly to the cost of a plant.

If multi-acre IBR hybrid plants are to be employed as generation assets to support blackstart, applying CIP-014 to those plants would present additional challenges. Yet, if large numbers of these plants are deployed over large areas, that dispersal may mitigate the risks of physical attacks in important ways. The widely distributed nature of such generation would complicate an adversary's attack planning and execution, and make it more difficult for adversaries to disrupt the same total amount of generation as they could be striking a single, very large scale next start plant. Thus, risk assessments and security requirements for dispersed IBR facilities may be different than those for conventional generating facilities.

C. Financial Incentives

The August 2023 Electric Reliability Organization Reliability Risk Priorities Report frames the overall problem for ensuring that restoration plans and capabilities remain sufficient as the grid evolves. The report notes that "transformation of the resource mix can alter the provision of and need for essential reliability services" and other prerequisites for reliability. The report also finds that to plan for future operations, "restoration services, such as blackstart capabilities and procedures, must also be considered. For services that may become scarce in the future, both organized markets and bilateral markets must anticipate and procure sufficient services to ensure reliable operations.[297]

297 NERC, RISC ERO Reliability Risk Priorities Report, August 17, 2023, 25, https://www.nerc.com/comm/RISC/Related%20Files%20DL/RISC_ERO_Priorities_Report_2023_Board_Approved_Aug_17_2023.pdf

Compensation constitutes a crucial component of such efforts. For the initial phase of blackstart restoration, PJM's tariff initiative for fuel assured blackstart units provides a valuable model of how additional compensation can incentivize GOs to provide adequate (and more resilient) resources. Additional funding options, including for capital investments to support deployment units where shortages are most severe, may also be effective for other regions and for U.S. security partners.

Solving the financial incentive problem on a more stable basis requires the development of standard engineering and planning methods that allow for the accurate estimation of financial risk associated with blackouts and black starts, and that link investments in grid resilience with the mitigation of those risks, so the investments can be optimized and built into the rate-setting process as adequacy and reliability investments are today. Such methods are in their infancy and therefore could benefit from research and development investment.[298] Such methods promise to facilitate investment in blackstart resilience by enabling accurate cost-effectiveness determinations that would enable the existing regulatory and rate-setting process to appropriately value blackstart resilience expenditures.

The owners of those assets in areas with organized markets typically receive no additional funding to contribute to blackstart restoration, even though such contributions are absolutely vital to growing power islands and maintaining their stability.

The situation for next start is more complicated. The owners of those assets in areas with organized markets typically receive no additional funding to contribute to blackstart restoration, even though such contributions are absolutely vital to growing power islands and maintaining their stability. For example, PJM Schedule 6A, Black Start Service, provides for recovery of new or additional capital costs and other sources of compensation for blackstart units, but not for next start assets.[299] The latter assets receive only normal cost-

298 Mouco, A., et al. A Discussion of Resilience in Power Grids; Proc. CIGRE U.S. National Committee 2023 Grid of the Future Symposium, 2023.

299 PJM, Schedule 6A, Black Start Service, https://www.pjm.com/directory/etariff/MasterTariffs/23TariffSe

based or market-based compensation even if they will play crucial roles in re-establishing the interconnections after a Black Sky event.

Longer term, a regulatory approval process must be developed that accounts for an operator's investments that benefit the resilience of the entire grid. New standard methods are required to form the basis for measuring the importance of each asset or facility for prevention of and recovery from widespread cascading failure events including blackstart.

Recommendation: Regulators and their industry partners should consider developing tariff mechanisms for next start assets in organized markets to receive compensation both for CIP compliance and other potential costs of supporting blackstart restoration. In particular, as Steve Naumann notes, "the elephant in the room is the risk of penalties for non-compliance. NERC penalties are not recoverable in rates. Also, for entities that own (or whose affiliates own) multiple generation facilities, a violation of a NERC Reliability Standard can affect penalties in the future on other assets."

Regulators and their industry partners should consider developing tariff mechanisms for next start assets in organized markets to receive compensation both for CIP compliance and other potential costs of supporting blackstart restoration.

ctions/4406.pdf

D. Employing DERs for Blackstart Versus BES-Connected Assets

In theory, TOPs could use either distribution, sub-transmission, or BES-level IBRs as supplemental sources of electric power for blackstart operations (or some combination of both – see the Distribution Grid Black Start (DGBS) discussion in Chapter VII). PJM's fuel-assured blackstart unit initiative explicitly proposes that DERs be allowed to compete to provide blackstart services along with BES-connected IBRs. In practice, DERs have significant disadvantages that may persist for years to come.

The Department of Energy notes that during the early phases of blackstart, the energized regions are very susceptible to disruption and collapse due to weak grid conditions. Using DER to support blackstart restoration in these early phases will be difficult "because these devices are mixed with load."

The Department of Energy notes that during the early phases of blackstart, the energized regions are very susceptible to disruption and collapse due to weak grid conditions. Using DER to support blackstart restoration in these early phases will be difficult "because these devices are mixed with load." In particular, "it would not be possible to energize a distribution system without bringing online uncontrolled local loads, such as air conditioning, refrigeration, and lighting, on the same distribution circuit.[300] Extensive use of DER for a blackstart would probably need to be coupled with developing networked or ad hoc microgrids combined with extensive controllable loads."[301]

300 In theory, with enough switches on a distribution feeder, many of the loads could be stripped off. However, that would take time as the distribution automation devices likely would need to go through a lengthy process to isolate different parts of a feeder. Moreover, T/D transformers have relatively high impedance compared to T/T transformers. Consequently, voltage drops, especially due to inrush current during starting of large motors, may limit the capability of DERs to start next start generators. Steve Naumann, note to the author, 8.16.2023.

301 DOE, Cybersecurity Considerations of Distributed Energy Resources, October 2022, 28, https://www.energy.gov/sites/default/files/2022-10/Cybersecurity%20Considerations%20for%20Distributed%20Energy%20Resources%20on%20the%20U.S.%20Electric%20Grid.pdf

Steve Nauman notes an additional problem posed by DERs such as rooftop solar assets. While not part of restoration plans, these assets can effectively function as negative loads; when a feeder is switched on, the load that the grid sees is net of load less local generation. Moreover, owners of these systems typically take the equipment as installed and do not check and test equipment settings as time goes on. If the controls on rooftop solar assets remain on after feeders get restored, they could trip off due to voltage and frequency fluctuations. This would result in load increases at the time the generation-load balance is fragile. The resulting instabilities would be especially significant for feeders with large amounts of local generation that are reconnected during the initial phases or restoration. Effective, nationwide compliance with IEEE 1547-2016 standards could mitigate such risks. Many owners of rooftop solar systems likely will not reset parameters from defaults or continue testing and resetting as needed. Comprehensive compliance with IEEE standards will take years to achieve, however.

Dealing with the problems of uncontrolled local loads will almost certainly present stability and control problems for any future restoration strategy, regardless of generation source. Nevertheless, in comparison with DERs, BES-connected IBRs have significant advantages. These assets are registered with NERC and subject to interconnection agreements with transmission providers.[302] These providers and their BES partners will face fewer challenges in accounting for and managing the flow of IBRs connected to their systems than distribution-level assets. Especially in the case of hybrid plants with advanced BESS inverters, such assets should largely replicate the functions (and challenges for managing island formation and growth) of conventional sources of power. That IBR functionality will not make blackstart restoration any easier, especially in the aftermath of Black Sky events. But nor should it make blackstart more difficult than it already is.

302 Interconnection agreements (IAs) are often made between Transmission Providers, which is a FERC-defined term versus one employed by NERC. In RTOs, the IA are made between both the Transmission Owner and Transmission Provider.

IV | EMPLOYING GRID-FOLLOWING AND GRID-FORMING IBRs FOR BLACKSTART

Above and beyond the ride-through and ERS-support capabilities of IBRs, advances in the ability of inverters to "form" the grid in the aftermath of power outages may also help these resources contribute to blackstart, and – potentially – enable the development of new restoration strategies that provide greater resilience against Black Sky threats.

DOE summarizes these opportunities. To restart the grid after a blackout in the current hybrid system, grid operators must first start up synchronous generators before they can connect IBR-generated power to the grid. The reason: when a large disturbance or outage occurs on the grid, conventional inverters will shut off power to these energy sources and wait for a signal from the rest of the grid that the disturbance has settled, and it is safe to restart—known as "grid-following." As wind and solar account for increasing shares of the overall electricity supply, it is becoming impractical to depend on the rest of the grid to manage disturbances. Grid-forming inverters are an emerging technology that allows solar and other inverter-based energy sources to restart the grid independently.[303]

TOPs in the U.S. and abroad are already conducting large-scale experiments with grid-following (GFL) inverters to help manage instabilities in IBR-heavy systems and prepare for their use to support restoration operations. But at very high levels of IBR penetration, grid operators are also experiencing reliability problems caused by the performance of these assets and their dependence on signals from the shrinking synchronous fleet.

As wind and solar account for increasing shares of the overall electricity supply, it is becoming impractical to depend on the rest of the grid to manage disturbances. Grid-forming inverters are an emerging technology that allows solar and other inverter-based energy sources to restart the grid independently.

Grid-forming inverters (GFM) provide an additional opportunity to help IBRs contribute to grid stability and blackstart restoration, and to also compensate for the limitations of their grid-following counterparts. The analysis below examines the technical differences and operational challenges of these two types of inverters, the ways in which they might support blackstart, and the implications for long-term restoration planning and system resilience.

303 DOE, Powering On with Grid-Forming Inverters, January 4, 2021, https://www.energy.gov/eere/solar/articles/powering-grid-forming-inverters

A. Grid-Following vs Grid Forming Inverters

Synchronous generators have long provided the capabilities needed to form the grid when major disturbances or blackouts occur. NREL notes that these generators "regulate their terminal voltages and respond to a change in grid frequency through a change in its power output; these are traditionally referred to as generator excitation and turbine-governor controls, respectively. These types of primary, secondary, and tertiary controls are well-known. We refer to these generation sources as grid-forming."[304] With the addition of IBRs to the grid, and state and federal goals for achieving zero-carbon generation, ensuring the availability of such grid-forming capabilities poses a critical challenge.

GFL inverters that are widely deployed today cannot provide these capabilities. Current inverter-based generation sources generally use phase-locked loops (PLLs), which rely on externally generated voltages by synchronous machines to operate. In case of unintended separation of the power system, islanded systems comprising only GFL inverters will not be capable of functioning autonomously. Similarly, after a blackout, GFL inverters cannot support the restoration process of the bulk power system unless synchronous generators with blackstart capability are present.[305]

GFM inverters are designed to replicate the functionality provided by synchronous assets. NERC notes that "there are presently no universally agreed upon definitions of GFL and GFM inverter controls in the industry," but offers the following definition:

> *Grid Forming Control for BES-Connected Inverter-Based Resources are controls with the primary objective of maintaining an internal voltage phasor that is constant or nearly constant in the sub-transient to transient time frame. This allows the IBR to immediately respond to changes in the external system and maintain IBR control stability during challenging network conditions. The voltage phasor must be controlled to maintain*

304 Yashen Lin et al, Research Roadmap on Grid-Forming Inverters, National Renewable Energy Laboratory (NREL), 2, November 2020, https://www.nrel.gov/docs/fy21osti/73476.pdf

305 Ibid

> *synchronism with other devices in the grid and must also regulate active and reactive power appropriately to support the grid.*[306]

In contrast to GFL, GFM most commonly uses an instantaneously measured voltage signal rather than a processed signal from a PLL as in a GFL inverter. GFM response and support to the grid are instantaneous. NERC states that GFM technology has been recently proposed in the IBR industry for use in parallel with the BES which, "when mature, is expected to address the majority of the risks and concerns of high (up to 100%) IBR penetration grid operation and stability with coordinated control and appropriate studies."[307]

GFM research and deployment efforts are accelerating worldwide. In the U.S., DOE is funding a consortium to advance the research, development, and commercialization of IBR grid-forming technologies to enhance power systems operation in August 2021. National Renewable Energy Laboratory's Research Roadmap on Grid-Forming Inverters report provides a roadmap on GFM controls, IBR's impact on grid stability, and the evaluation of crucial system interactions for the future grid that will be comprised of the combination of synchronous generators and GFM as well as GFL inverter-based resources.

National Renewable Energy Laboratory's Research Roadmap on Grid-Forming Inverters report provides a roadmap on GFM controls, IBR's impact on grid stability, and the evaluation of crucial system interactions for the future grid that will be comprised of the combination of synchronous generators and GFM as well as GFL inverter-based resources.

Hawaiian Electric Company plans to implement wide-spread GFM BESS technology throughout their island power systems by 2023. Significant work in specification, procurement, and detailed Electromagnetic Transient (EMT)

306 NERC, White Paper: Grid Forming Technology: Bulk Power System Reliability Considerations, December 2021, 8-9, https://www.nerc.com/comm/RSTC_Reliability_Guidelines/White_Paper_Grid_Forming_Technology.pdf

307 Ibid

studies have been completed. These detailed studies indicate that the GFM technology will be critical to allow the envisioned high renewable penetration scenarios to operate reliably.[308]

Great Britain is developing the grid code modification GC0137 to add requirements for GFM IBRs and high voltage direct current systems to ensure stable and reliable operation of the grid with high penetration of IBRs. Australia has been actively developing strategies to use GFM IBRs to support high IBR penetration into the power system. For example, recent research has shown that GFM IBRs provide a successful alternative to synchronous condensers that have traditionally been used to provide system strength and voltage support. Australia continues to explore numerous options for increasing system strength.

Despite this progress, however, many technical and operational issues remain to be resolved as GFM begin to supplement (and, increasingly, supplant) grid following resources.

1. System Strength and GFL/GFM Integration in Hybrid Systems

The majority of the inverters used today in wind, solar, and energy storage resources are GFL. They read the voltage and frequency of the grid, lock onto it, and inject power aligned with that signal. However, instability can result in areas with high levels of GFL inverter-based resources relative to conventional synchronous generators such as coal- and natural gas–fired plants and hydroelectric plants. The reason for this instability is that the voltage signal is easily perturbed as GFL IBRs are injecting power into the grid, making it harder for them to lock onto the grid voltage signal correctly. Moreover, the conventional synchronous generators

Interconnection queues in the United States contain more capacity than the total currently installed generation capacity, with the vast majority of projects being wind, solar, and battery storage. In the absence of incentives or requirements for control improvements, all of the IBRs built (likely 20 to 30 percent of those in the queue) will have conventional GFL controls, thus exacerbating stability concerns.

308 CIGRE

that provide strong voltage signals tend to be located far from areas rich in renewable resources. The greater the distance between pockets of renewables and conventional synchronous generation, the weaker the voltage signal from those strong voltage sources.

These challenges are getting more severe as remote sun- and wind-rich areas attract continued development of GFL IBRs, as seen, for example, in West Texas and New England in the United States, in South Australia, and in many other regions. Interconnection queues in the United States contain more capacity than the total currently installed generation capacity, with the vast majority of projects being wind, solar, and battery storage. In the absence of incentives or requirements for control improvements, all of the IBRs built (likely 20 to 30 percent of those in the queue) will have conventional GFL controls, thus exacerbating stability concerns. [309]

Another challenge for future high penetration inverter-based resource systems is how the grid forming inverters synchronize with each other. Grid forming inverters are able to start up on their own without support from the grid. Grid following inverters track the grid that could be formed by the grid forming inverter. Under these very low short circuit strength conditions, special care may be needed to ensure stable operation between the grid following inverters. Additional grid forming inverters may be added to the BES. However, they need to ensure that their output voltage magnitude and phase at the inverter are the same as the grid-side of the inverter circuit breaker where it connects for proper synchronization. Grid forming inverters could use a PLL to track the grid in grid supportive mode for a short time, synchronize, and then return to grid forming mode once synchronized. Future work is needed in this area to develop operating plans.

2. Additional Consideration for Employing GFM IBRs as Blackstart Units

NERC notes that GFM technology provides a significant advantage over conventional GFL technology because GFM inverters can be designed as an

309 Julia Matevosyan, Energy Systems Integration Group, Capturing the Benefits of Grid-Forming Batteries: A Unique Window of Opportunity, February 6, 2023, https://www.esig.energy/capturing-the-benefits-of-grid-forming-batteries-a-unique-window-of-opportunity/

initial blackstart resource that is not reliant on a strong synchronous grid to reestablish the grid from an outage. Depending on the location of the GFM IBRs in the network, IBRs can serve as a traditional, centralized blackstart resource or as a distributed black start resource if located at the edge of the grid. Each TOP and RC developing a restoration plan and studying cranking paths should fully understand how any potential GFM IBRs being used in the path would operate. EMT studies are needed to fully understand how the GFM resource will operate under such low system-strength conditions and to validate that the GFM IBR can energize transformers, charge lines, and start motor loads. GFM technology may be able to help shorten outage durations and provide grid resilience; however, energy assurance from variable weather and inverter overcurrent capability will also need to be considered.[310]

- **Recommendation:** dispatchable inverter-based resources could be capable of operating as a blackstart resource, when utilizing inverters operating in grid forming mode, since they are able to generate their own voltage waveform and are not reliant on the BES to synchronize. Blackstart service from BES-connected inverter-based resources should be implemented in coordination with the TOP, RC, BA, etc. The implications and impacts these controls could have on synchronous machines also need to be more clearly understood, particularly during the transition to a very high penetration inverter-based power system.[311]

Dispatchable inverter-based resources could be capable of operating as a blackstart resource, when utilizing inverters operating in grid forming mode, since they are able to generate their own voltage waveform and are not reliant on the BES to synchronize.

It will also be crucial to ensure that as grid-forming technologies advance, these capabilities are applied to the deployment of BESS – especially for use as blackstart assets. While GFM controls can

310 NERC, Grid Forming Technology, 16

311 NERC | Inverter-Based Resource Performance Guideline, September 2018, 65-66, https://www.nerc.com/comm/RSTC_Reliability_Guidelines/Inverter-Based_Resource_Performance_Guideline.pdf

> potentially be implemented on any type of IBR including new solar and wind plants, batteries are the low-hanging fruit. GFM behavior requires a certain amount of energy buffer, which for wind and solar resources means they must continuously operate below their maximum available power production. In addition, GFM control in wind turbines may result in greater and more frequent mechanical stress. But utility scale batteries have none of those drawbacks. The battery is the energy buffer, and only software modifications to a battery's controls are needed to make the battery a GFM resource.

In late 2022, the Australian Renewable Energy Agency announced co-funding of eight large-scale GFM batteries across Australia with a total capacity of 2 GW/4.2 GWh, to be operational by 2025. In Great Britain five new large grid-connected GFM batteries will be deployed between 2024 and 2026. Large equipment manufacturers such as SMA, Tesla, and Hitachi already have commercial offerings of GFM controls in BESS.

While only a relatively small number of utility-scale batteries are installed in the U.S. today, a significant amount of battery capacity will likely be developed in the next few years. Without specifications and the appropriate incentives or requirements, much or all of this capacity will likely lack GFM capability, which would result in continued stability challenges, continued solar and wind curtailment, and the need for costly supplemental stabilizing equipment such as synchronous condensers or new transmission build-out. In contrast, with specifications and the appropriate incentives or requirements, this battery storage capacity can be installed with GFM capability and lead to improved grid stability, less curtailment, and little or no need for additional costly stabilizing equipment.[312]

with specifications and the appropriate incentives or requirements, this battery storage capacity can be installed with GFM capability and lead to improved grid stability, less curtailment, and little or no need for additional costly stabilizing equipment.

312 Julia Matevosyan, Energy Systems Integration Group, Capturing the Benefits of Grid-Forming Batteries: A Unique Window of Opportunity, February 6, 2023, https://www.esig.energy/capturing-the-benefits-of-grid-forming-batteries-a-unique-window-of-opportunity/

B. Specifications for GFM Inverters: Making Survivability a New Priority

A DOE-supported initiative is underway to develop universal performance standards for grid-forming inverters. The Universal Interoperability for Grid-Forming Inverters (UNIFI) Consortium has developed a preliminary version of specifications for these devices. The Consortium's goal is "to provide uniform technical requirements for the interconnection, integration, and interoperability of GFM IBRs of any size in electric power systems of any scale."[313]

Version 1 of these specifications offers a range of performance requirements for operation in normal and abnormal electric system conditions, including ride-through capabilities and frequency response. Many of these same capacities are addressed in IEEE Standard 1547. In addition, however, the Consortium proposes capabilities to enable IBR support for blackstart and system restoration. The Consortium notes that not all GFM IBRs are expected to be able to provide black-start services, since those services may require additional hardware, design, and functionalities. Accordingly, making IBRs capable of blackstart is likely to incur additional cost and require special coordination with system operators. That said, the Consortium proposes a number of generic performance requirements for GFM IBRs that are designated to provide blackstart services. These devices should be able to:

1. Provide a soft start to loads (i.e., by ramping up the output voltage at a programmable rate) which would alleviate inrush currents.
2. Supply transformer, cold-load, and other inrush current needed. This process is a key step in black start since it may cause significant inrush current, especially from transformers and motor loads, which is generally more challenging when using IBRs than synchronous generators, due to lower over-current capability.
3. Provide a steady voltage reference for GFM and GFL IBRs and other resources to synchronize to and facilitate a smooth system restoration. Note some GFM IBRs may be configured to synchronize their own

313 B. Kroposki et. al, UNIFI Specifications for Grid-forming Inverter-based Resources— Version 1, UNIFI-2022-2-1, December 2022, 6

voltage and phase angle to a primary GFM that is providing the voltage waveform for the entire system.

4. Operate in parallel with other black-start GFM IBRs to achieve collective black start. This is optional and only needed for black start using multiple GFM IBRs. This technique is used when the baseline load exceeds one single GFM IBR's capability. This functionality may incur additional complexity, but it would increase flexibility and survivability of the system.[314]

An August 2023 report by CIGRE, Power System Restoration Accounting for a Rapidly Changing Power System and Generation Mix, offers additional recommendations on using grid-forming BESS not only as initial blackstart resources but also for broader restoration functions. In particular, the report recommends that to utilize grid-forming inverters as stand-alone blackstart assets, a soft energization (or soft start) mechanism is required within inverter controls to allow for gradual voltage increment (ramp up), where the inrush current can be limited.[315]

The report also notes that significant performance issues remain to be resolved before GFM IBRs can be employed for blackstart. Undervoltage protection issues are especially significant. During the restoration process, multiple feeders could drop out when the voltage is below certain thresholds, so customized protection settings will be needed for the duration of start-up. CIGRE proposes a number of measures and possible standards to mitigate these and other problems for utilizing GFM IBRs for blackstart.

The benefits of establishing common GFM performance requirements are immense. Rather than deploying these resources across the grid with varying performance features, different grid-forming capabilities, and device-specific requirements for integration into blackstart operations, having a widely shared set of specifications will be enormously helpful.

314 UNIFI Specifications, 13

315 CIGRE, Power System Restoration Accounting for a Rapidly Changing Power System and Generation Mix,

Yet, establishing these standards is not a panacea for blackstart resilience. No mechanisms yet exist to enforce the adoption of such standards by device manufactures or require their purchase by utilities instead of buying lower cost, less capable IBRs.

- **Recommendation:** Specialized cost recovery measures and other incentives may be necessary to support the deployment of blackstart-capable GFMs.

Moreover, given the operating environment in which GFM inverters will need to conduct restoration, much more attention should be paid to their survivability. The Consortium states that "cybersecurity requirements can introduce delays and challenges" in developing and implementing IBR performance options for real-time control and power system stabilization.[316] Yet, unless cybersecurity is engineered into these devices, their ability to help restore and stabilize the grid in the event of a cyber conflict with China or other adversaries will be dubious at best. The same is true of resource survivability against EMP effects.

- **Recommendation:** Measures to develop and implement performance standards for GFM IBRs must go hand in hand with measures to ensure their resilience in Black Sky events.

The CIGRE report offers a useful assessment of the basic cybersecurity challenges that restoration will entail. A key finding: "currently a holistic view from the perspective of resilience is missing." First, however, it will be helpful to highlight the "big picture" resilience opportunities that IBRs may be able to provide in combination with other types of resources.

Measures to develop and implement performance standards for GFM IBRs must go hand in hand with measures to ensure their resilience in Black Sky events.

316 UNIFI Specifications, 13

V | DECENTRALIZATION AND GENERATION DIVERSITY: **IBR-ENABLED STRATEGIES FOR BLACKSTART RESILIENCE**

Current blackstart plans rely on a relatively small number of blackstart units that energize an equally small number of cranking paths. Significant opportunities exist to strengthen the resilience of this infrastructure against Black Sky hazards, and to facilitate such measures through new regulatory and cost recovery initiatives. IBRs with grid-forming capabilities could enable TOPs and their partners to pursue a very different strategy. Instead of hardening a limited set of assets, transmission operators could deploy hundreds of GFM-capable resources to establish multiple power islands.

An asset dispersal strategy of this type will only be practical if the myriad technical and operational problems discussed above can be resolved. Moreover, widely dispersing grid-forming IBRs will fail to strengthen resilience if those inverters are subject to common mode failures. DOE's report on Cybersecurity Considerations for Distributed Energy Resources identifies a number of vulnerabilities that could be widely shared by IBRs.[317] Countering these vulnerabilities will be essential to avoid the nationwide disruption of GFM-led restoration operations.

Another way of mitigating the risks of such widespread inverter failure is to maintain diversity in the resources dedicated to blackstart. In particular, by retaining natural gas-fired generators to support grid-forming operations in a heavily decarbonized grid, we could force adversaries to develop and execute very different types of attacks to cripple blackstart restoration.

Current blackstart plans rely on a relatively small number of blackstart units that energize an equally small number of cranking paths.

Basing resilience on both generation diversity and dispersion also makes a virtue out of necessity. Even with the significant progress that is underway in developing and deploying GFMs, we are many, many years away from being able to rely exclusively on those systems for blackstart. Gas-fired generation will likely remain essential for power restoration (and especially for next start) for a decade or more. More specifically, the U.S. and its partners will rely on hybrid systems that combine voltage-setting conventional generation – especially natural gas – with a growing number of inverter-based resources, including both grid-following and grid-forming assets. That is why ensuring the resilience of fuel supplies for gas generation is a vital national security requirement.

317 DOE, Cybersecurity Considerations, 20-29

7 CHAPTER SEVEN

ADDITIONAL GENERATION OPTIONS FOR BLACKSTART OPERATIONS

In this chapter, we review non-traditional generation options for blackstart operations. Among these are Small Hydropower Generation, Small Modular Nuclear Reactors, Geothermal Energy, and Fuel Cells. We also discuss the important concept of multi-level blackstart, in addition to the BES transmission-level paradigm.

I | SMALL HYDROPOWER GENERATION FOR BLACKSTART AND SYSTEM RESILIENCE

Large hydropower generators are broadly recognized as the premier blackstart generation resource. They meet all the requirements for blackstart, as outlined in Section 1.A, "Blackstart 101". They do not require external grid power to start and operate; require minimal "house power" to operate control systems; are able to (re)start quickly; provide both real and reactive power; they have nearly unlimited "fuel" in their reservoirs (though this can be affected by droughts or seasonal variations); can deal with grid frequency swings and provide frequency stabilization; and they typically have direct transmission connections to other "next start" generators.[318]

Given these characteristics, large (>10MW) hydropower generators have long been considered the "gold standard" for BES // transmission-level black start. Smaller hydroelectric plants (100 kW-10 MW) are fundamentally the same, and so possess nearly all of the blackstart characteristics listed above and present an opportunity to dramatically increase the available number of blackstart generators.[319] They also represent the basis for a new paradigm of distribution grid blackstart (DGBS) operations, one that could also be applied to other (smaller, renewable) generation sources. Though the bulk of this report focuses on BES / transmission level blackstart, in the interest of true Black Sky resilience, DGBS offers an opportunity to broaden blackstart capability and enhance overall (generation, transmission, sub-transmission, distribution, and microgrid) resilience. Figure 7.1 below shows the locations of U.S. hydropower assets of all sizes including all powered dams.[320]

large (>10MW) hydropower generators have long been considered the "gold standard" for BES // transmission-level black start.

Smaller hydroelectric plants (100 kW-10 MW) are fundamentally the same, and so possess nearly all of the blackstart characteristics listed above and present an opportunity to dramatically increase the available number of blackstart generators.

Throughout the U.S. and Europe, legacy infrastructure from various mills that manufactured clothing, processed food, and produced other products using waterpower are still in existence. Many of these facilities were converted

318 DOE HydroWires; Hydropower Plants as Black Start Resources, www.energy.gov/eere/water/articles/hydropower-plants-black-start-resources

319 IEEE; Kurup and Ashok, Performance of a Hydropower Plant During Black Start and Islanded Operation; https://www.scribd.com/document/353763378/Hydro-Power-Plant

320 Oak Ridge National Laboratory, The 2020 National Hydropower Map, https://hydrosource.ornl.gov/map/2020-national-hydropower-map

from water-to-mechanical power to water-to-electric power in the early 1900s.[321] Most of these former mills have since closed, but the small dams, hydropower generators, and surrounding infrastructure are still in place and are either currently operational or can be made so with minimal investment.

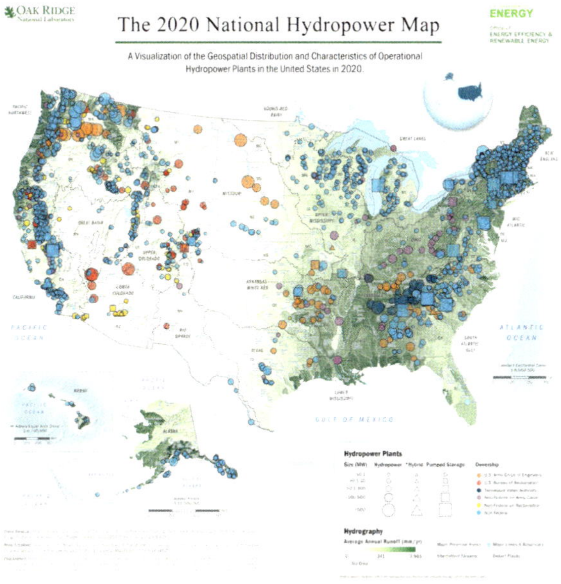

Figure 7.1, The 2020 National Hydropower Map.

In addition to dams built for industry, the expansion of agriculture in the United States during the 18th and 19th centuries was founded in part on the construction of dams to provide irrigation and drainage. Due to the rural nature and small energy potential of many of these developments, the hydroelectric resources were not developed when the dams were originally installed.[322] Retrofitting of these existing dams to produce energy, in addition to their primary use, can be made with significantly less investment than new

321 Compus Rendus Mechanique; Pierre-Louis Viollet, From the Water Wheel to Turbines and Hydroelectricity, Technological Evolution and Revolutions, https://www.sciencedirect.com/science/article/pii/S163107211730092X

322 Ambio; Baldassarre, Mazzoleni, Rusca, The Legacy of Large Dams in the United States https://www.ncbi.nlm.nih.gov/pmc/articles/PMC8363679/

greenfield developments. In addition, many of the permitting and licensing issues involving the construction of new dams can be avoided.[323]

There are approximately 90,000 dams in the U.S. Currently, only 2,200 are powered. A map of all dams, powered and unpowered, is shown in Figure 7.2 below,[324] which illustrates the distributed nature and the magnitude of the existing infrastructure.

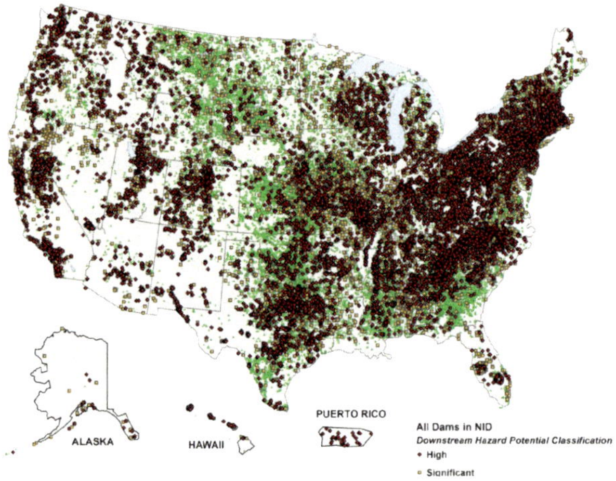

Figure 7.2, Location of all powered and non-powered dams in the U.S..

Most of the existing powered dams are greater than 1 MW and appropriate for blackstart operations.[325] Many of the existing small generators are co-located with U.S. critical infrastructure, see figure 7.3.[326] In addition, the concentration of non-powered dams in the eastern U.S. overlaps many areas of dense critical infrastructure.

323 Yale Environment 360; James Dinnen, Can Retrofitting Dams for Hydro Provide a Green Energy Boost? https://e360.yale.edu/features/can-retrofitting-dams-for-hydro-provide-a-green-energy-boost

324 FEMA, Dam Safety in the United States, A Progress Report on the National Dam Safety Program, January 24, 2014, https://www.fema.gov/emergency-managers/risk-management/dam-safety/progress-report

325 PNNL, Electric Grid Blackstart: Trends, Challenges, and Opportunities, 2022,. www.pnnl.gov/main/publications/external/technical_reports/PNNL-32773.pdf

326 DOE OSTI; Enhancing Local Grid Resilience with Small Hydropower Hybrids: Proving the concept through demonstration, simulation, and analysis with Idaho Falls Power; https://www.osti.gov/biblio/1891110

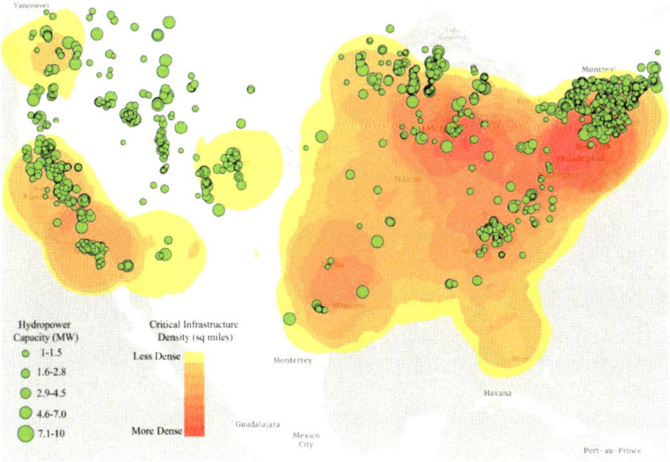

Figure 7.3, Small Hydropower plants and their proximity to critical infrastructure, in the U.S..

Based on the U.S. Army Corps Engineers National Inventory of Dams (USACE NID), many of these non-powered dams have small drainage basins or have a low height that led to small available capacities. There are approximately 20,000 non-powered dams in the USACE NID that are greater than 10 ft in height and have a drainage area of more than 10 mi2. [327]

The growing interest in clean, renewable energy is driving a renewed interest in using these systems for local power generation. Beyond the ability to provide power to the BES, local load possibilities for the small hydro generators are EV charging stations, specific facilities like industrial, data centers, hospitals, police, fire, or providing power directly into microgrids or local distribution systems. Critical loads, powered by local small hydropower, can also be an important concept for Black Sky resilience. Ensuring that critical loads are more resilient to BES disruptions enhances the overall resilience of the electrical system and society writ large, by decreasing the overall load that must be quickly restored by the BES in the event of a Black Sky event.

327 USACE; National Inventory of Dams, https://nid.sec.usace.army.mil/#/

II | OTHER POSSIBLE BLACKSTART CRANKING PATHS SCENARIOS: **BEYOND BES-LEVEL GENERATION**

In a recent report, the Idaho National Lab and DOE describe possible blackstart cranking path scenarios that differ from the traditional path of starting from generators on the high-voltage transmission system (read: BES), which they term the "top-down" restoration system. They identify "four main categories of blackstart restoration strategies:"[328]

- **Top-down restoration with high-voltage transmission system**: In this traditional approach, a few central blackstart units are restored

328 Idaho National Laboratory; Enhancing Local Grid Resilience with Small Hydropower Hybrids: Proving the concept through demonstration, simulation, and analysis with Idaho Falls Power; https://www.osti.gov/biblio/1891110

first. These units then energize the high-voltage transmission system. Next the sub-transmission network to non-blackstart [next start] units is energized to establish the cranking path and eventually restore distribution grid and load services. Hydropower has long been a key transmission-level blackstart resource in many locations within the United States.

- **Bottom-up restoration with a sub-transmission system:** In this strategy, the blackstart units first energize the sub-transmission network and establish cranking path to larger but non-blackstart generation plants [next start units]. A parallel multi-island restoration is executed, and then, the high-voltage transmission system is energized on the way to entire grid restoration.
- **Bottom-up restoration with a distribution system:** Referred to as distribution grid black start (DGBS), this approach has the potential to leverage DERs, adaptive protection settings, and smart switching capability to restore part or all of a distribution grid. This approach is independent of transmission system restoration status. Furthermore, this is consistent with the concept of a microgrid that has all required grid capabilities and can connect or disconnect from the broader grid. [M]any small hydropower plants exist in proximity of the nation's critical infrastructure, but most are not used as blackstart resources.
- **Combined restoration:** Both the top-down transmission side and bottom-up distribution side restoration are executed in parallel; stability of respective frequency and voltage profiles are ensured and finally synchronized at the point of interconnection.

Figure 7.4 below, drawn from the same report, depicts each of the four strategies.

Under the current "top-down" paradigm, the Regional Transmission Operator (RTO) or Balancing Authority (BA) is responsible for balancing generation and load during normal operation and of course during blackstart operations. They are also responsible for ensuring adequate blackstart generation assets within their transmission network. In this traditional paradigm, sub-transmission and distribution systems do not have or deploy blackstart generation assets; all power comes from the transmission system (BES) level and flows "downward".

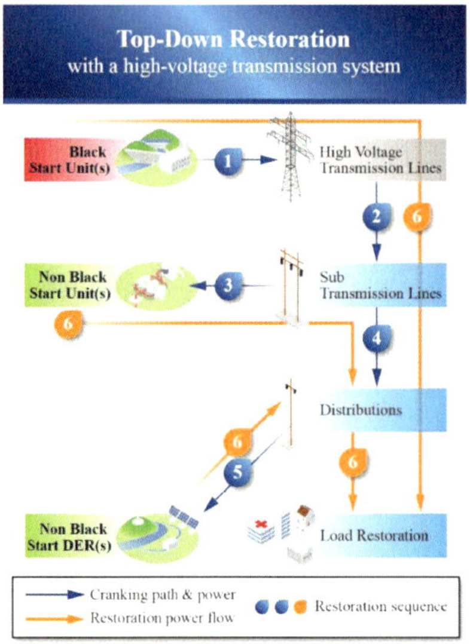

Figure 7.4.a Transmission Level Blackstart Restoration.

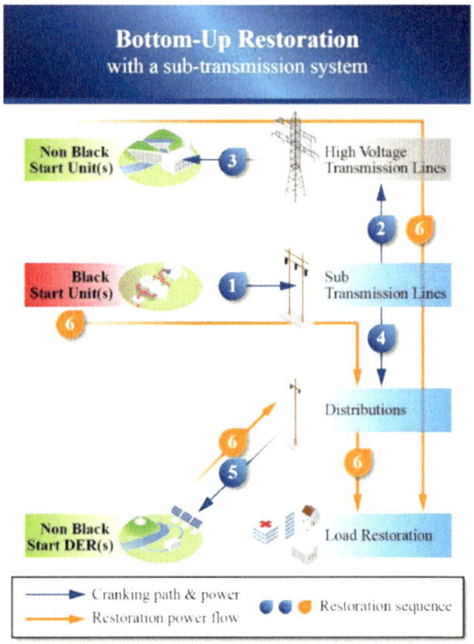

Figure 7.4.b Sub-Transmission Level Blackstart Restoration

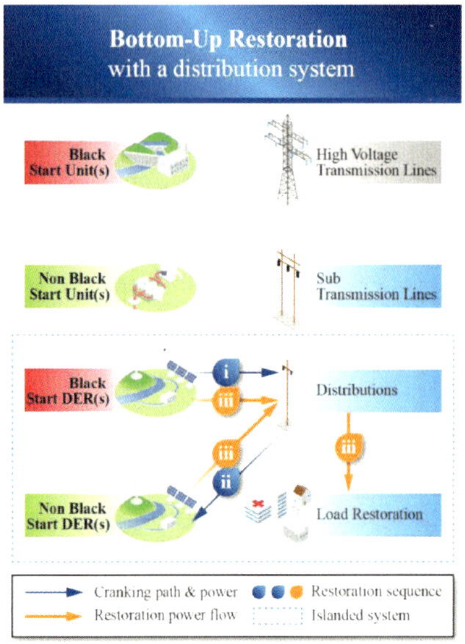

Figure 7.4.c Distribution Level Blackstart Restoration

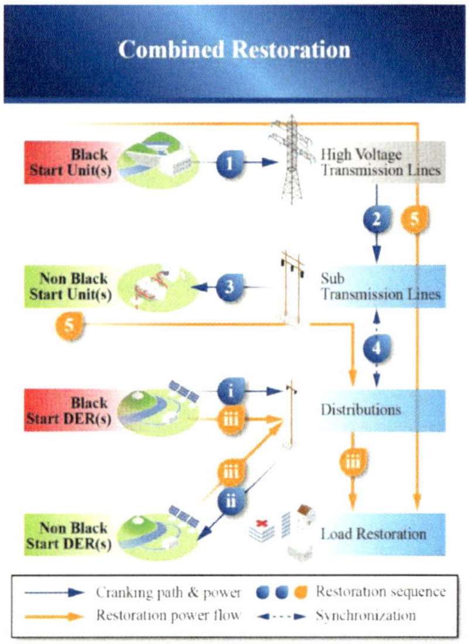

Figure 7.4.d Combined Blackstart Restoration

As the generation mix has grown to include new, green, distributed energy resources (DERs), other options beyond the top-down paradigm are now feasible. As shown in Figure 7.4, "bottom-up" blackstart restoration scenarios are possible that employ generation at the distribution or sub-transmission levels, or in combination. Supplementing the existing "top-down" paradigm for blackstart with coordinated use of "bottom up" approaches, combined with energy storage, can serve to enhance overall system resilience. Methods now exist to identify the optimal locations on the grid for the placement of bottom-up restoration investments; these locations tend to be critical nodes on the network topology and tend to participate in a large number of cascading failure pathways and blackstart cranking pathways.

One important caveat is that due to their smaller size, and therefore lower rotational inertia, coupling with an energy storage solution is required to better cope with frequency excursions. In their report Idaho National Lab (INL), in partnership with Idaho Falls Power (IFP), successfully demonstrated (though field tests and simulations) blackstart capability of the small IFP hydro units in combination with energy storage provided by ultracapacitors (field test) and battery backup (simulation).[329]

Recommendation: Beyond large hydropower units, smaller hydropower units can provide additional blackstart capability, if combined with energy storage, and "bottom up" (sub-transmission, distribution) restoration strategies. Thousands of small hydro generators are already in place in the U.S. and could be used to provide additional blackstart capability and overall electric system resilience at all levels: transmission, sub-transmission, distribution, and combinations of these. Additionally, thousands of existing non-powered dams distributed across the U.S. could be repowered and utilized to support distributed electric system resilience.

Note: The alternative blackstart generation cranking path options presented above could also apply to other, smaller generation sources besides hydropower, detailed below (advanced nuclear, geothermal, and fuel cells or batteries) and also for solar and wind IBRs discussed in previous chapters.

329 Ibid

III | GEOTHERMAL GENERATION FOR BLACKSTART AND SYSTEM RESILIENCE

The surface of our planet's core radiates at a temperature similar to our sun, driven by natural radioactive sources. Though there are many layers between the core and Earth's surface, heat from the core can leak up through cracks in the planet's crust. The most common forms of this phenomenon are volcanoes and geysers. These natural heat sources can be harnessed to generate power, known as geothermal energy. As in coal or nuclear power plants, steam is created from water being heated by (in this case) the geothermal source that then drives a steam turbine to produce electric power.

'The Geysers' in California's Sonoma and Lake counties is the largest geothermal field in the world. The Geysers' geothermal energy power generation is a collection of 22 units, ranging in size from 119 MW down to 20 MW, with a combined power output of over 1500 MW, accounting for 20%

of California's renewable energy generation.[330] Geothermal energy plants in the U.S. are currently located on the West coast, as well as Hawaii and Alaska, with California currently being the most prominent geothermal energy producing state in the U.S..

There are 31 geothermal power plants in the U.S., ranging in size from The Geysers at 1590 MW down to Lightning Dock in NM, at 4 MW.[331]

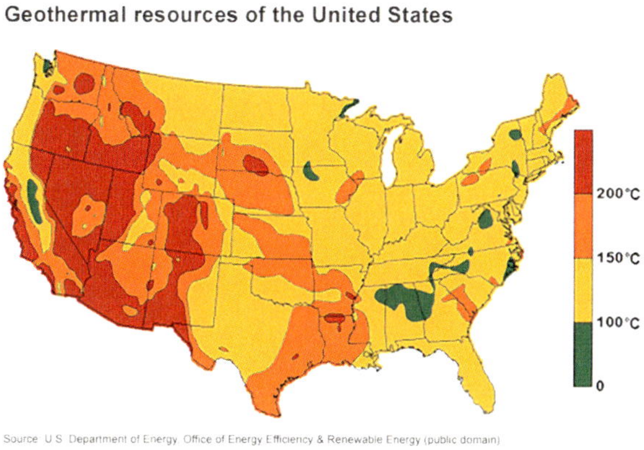

Figure 7.5, Map of Geothermal Energy Resources in the U.S.[332]

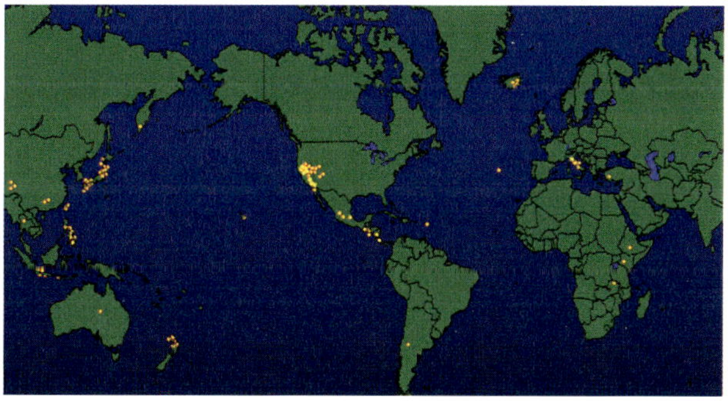

Figure 7.6, Geothermal Power Plant locations throughout the world.[333]

330 https://en.wikipedia.org/wiki/The_Geysers

331 https://en.wikipedia.org/wiki/List_of_geothermal_power_stations_in_the_United_States

332 U.S. Energy Information Administration, Geothermal Explained, https://www.eia.gov/energyexplained/geothermal/where-geothermal-energy-is-found.php

333 Geothermal Education Office; https://geothermaleducation.org/GEOpresentation/sld069.htm

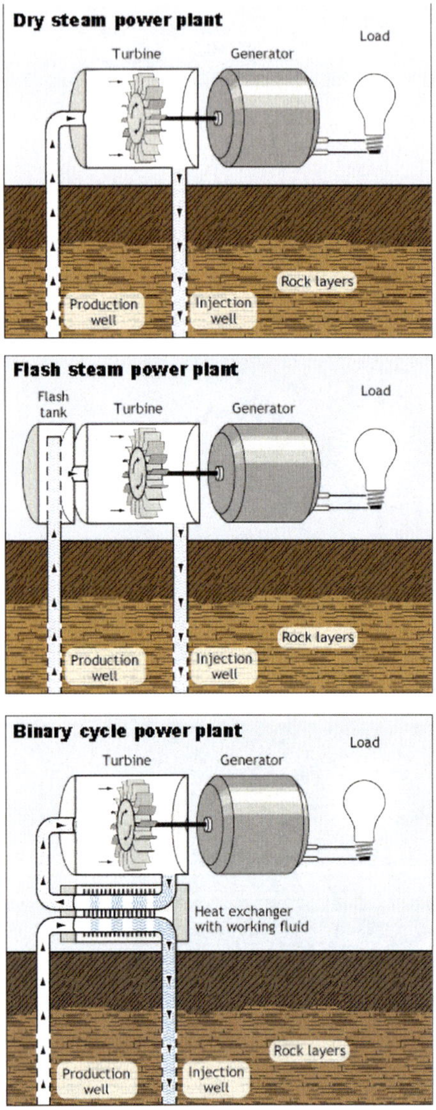

Geothermal energy can be harnessed from natural heat vents like geysers and volcanoes, or alternatively, heat vents can be drilled to harness geothermal energy in underground hotspot areas. Locations with temperatures of 250 degrees Celsius and above are the most viable hotspots for energy generation. Drilling a well allows for a steam turbine to be built above the well to convert the steam rising from the water heated below into power.

There are three basic types of geothermal power plants:[334]

- Dry steam plants use natural steam rising directly from a geothermal reservoir to drive generator turbines.
- Flash steam plants use high-pressure hot water from deep inside the earth and convert it to steam at the surface. The steam then drives generator turbines. When the steam cools and condenses, the water is injected back into geothermal reservoir to be reheated and reused.

334 U.S. Energy Information Administration, Geothermal Explained, https://www.eia.gov/energyexplained/geothermal/geothermal-power-plants.php#:~:text=Geothermal%20power%20plants%20require%20high,hot%20water%20to%20the%20surface.

- Binary-cycle power plants run the pressurized hot water coming from the geothermal reservoir through a heat exchanger to transfer the heat from the geothermal hot water to another working fluid [usually isobutene or isopentane].[335] The working fluid then turns to steam which drives a generator turbine. The working fluid recondenses, and it is reused for steam, and the geothermal water is sent back to the reservoir. Since the binary-cycle plant uses two closed fluid loops, it has virtually no emissions.[336]

Geothermal energy sources could be a useful blackstart asset. Unlike coal or nuclear plants, geothermal plants take very little energy to start up. The Theistareykir geothermal power plant in Iceland, for example, is capable of black start operations. On-site diesel generators provide the power to start the turbine to energize the GSU transformers and send power to the grid. Testing at the Theistareykir facility has also demonstrated that it can contribute positively to grid frequency regulation and that: "The response time of the units is significantly faster than a similar sized hydro power plant."[337] This frequency regulation capability is also critical in blackstart operations.

Geothermal energy sources could be a useful blackstart asset. Unlike coal or nuclear plants, geothermal plants take very little energy to start up.

As a Black Sky blackstart asset, geothermal energy offers very high energy resilience, especially when compared to other renewables like solar and wind technologies. If the sun isn't shining or the wind isn't blowing, no solar or wind energy is generated, whereas geothermal energy provides constant sustainable power, regardless of day/night or weather conditions.[338] Geothermal plants have virtually limitless heat/water/steam availability, and therefore no

335 DOE, Federal Energy Management Program; Geothermal Electric Technology, https://www.wbdg.org/resources/geothermal-electric-technology

336 Ibid

337 GRC Transactions; Hardarson et al, Theistareykir Geothermal Power Plant, Challenges in a Weak Electric Grid, https://publications.mygeoenergynow.org/grc/1034049.pdf

338 DOE EERE; Geothermal Basics; https://www.energy.gov/eere/geothermal/geothermal-basics

refueling requirement, an important aspect of Black Sky blackstart and next-start capabilities.

While the viable geographic locations of geothermal energy production are limited, the plants that do exist are - in principle - all capable of blackstart operations. Advances in drilling technology, coupled with the desire for renewable electricity generation, will allow for development of additional geothermal power plants in the U.S. and throughout the world.[339] One current example: the Secretary of the Air Force's Office Of Energy Assurance (SAF/OEA) is sponsoring two planned geothermal power pilot projects, one at Joint Base San Antonio, and the other at Mountain Home Air Force Base.[340] Both are currently somewhat outside of the "prime" geothermal energy locations (see Figure 7.5 above), but are now possible due to new techniques, as shown in figure 7.7 below.

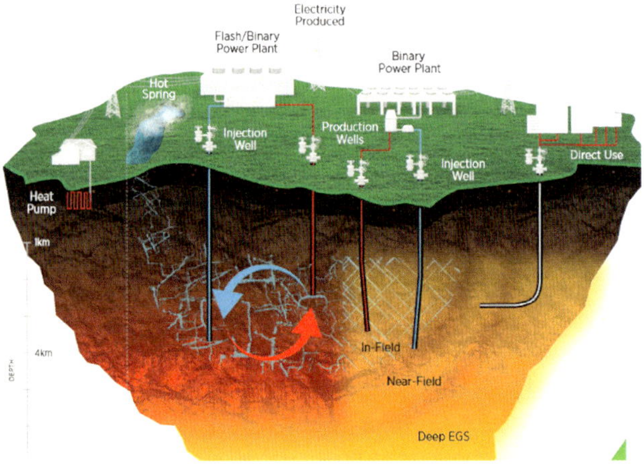

Figure 7.7: Geothermal Plant Prototype for SAF/OEA[341]

Recommendation: existing and planned geothermal plants of sufficient size and viable locations should be included as resources in blackstart planning.

339 Texas Geothermal Alliance, Could Geothermal Be the Energy Source of the Future? Texas Oil Companies Think, Maybe So; https://www.txgea.org/could-geothermal-be-the-energy-of-the-future-texas-oil-companies-think-maybe-so/

340 San Antonio Electromagnetic Defense Initiative, Quarterly Meeting, August, 18, 2023.

341 Ibid

IV | NUCLEAR POWER: FLEXIBLE OPERATIONS AND SMALL MODUALR REACTORS (SMR)

A. Motivation and Overview

The current, legacy fleet of nuclear power plants (NPPs) in the U.S. have been running reliably for more than 50 years. The existing U.S. nuclear power plants have continued running through Summer and Winter storms when Solar, Wind, Natural Gas and Coal plants have had to shut down. The chief reason is their robust safety design allowing for continued operation in the face of extreme weather conditions and other external threats.

However, if a sizeable Bulk Electric System (BES) gird disruption / blackout occurs wherein the NPP cannot draw power from the BES (usually called a "loss of outside power, or LOOP event), in accordance with current U.S. NPP standard operating procedures (SOP) and Nuclear Regulatory Commission (NRC) regulations, the NPPs will go offline and into "safe shutdown mode". Only when the BES is recovered and stable, do the nuclear power plants (slowly) restart and reconnect to the BES, and this restart requires outside electric power.

Given the Black Sky power outage operational challenges for all entities and sectors, NPPs going offline and moving to "safe shutdown mode" could lead to precarious situations. While in safe shutdown mode, the nuclear fuel does not stop producing heat for several months, which must be removed by active pumping[342]. Currently, emergency backup power to run active cooling, control, and safety systems of NPPs is supplied chiefly by diesel generators when they go offline and move into safe shutdown mode. For NPPs in the U.S., the Nuclear Regulatory Commission (NRC) Regulatory Guide 1.137 states that:

> *"All safety-related emergency power supply fuel oil systems should include the full 7-day supply in the fuel oil supply tank and should assume that the supply cannot be replenished for at least 7 days following the design basis event. The 7-day allowance assumes sufficient offsite supplies are available for continuous replenishment for a mission time of at least 30 days credited in the plant safety analyses. Pre-disaster agreements for fuel oil replenishment should be established in the event a natural calamity invokes widespread fuel shortages or results in extended outage of the transmission network."*[343]

This requirement has proven effective in all previous natural disaster scenarios or other LOOP events. In contrast to prior experience, however, Black Sky power outages could last from weeks to months and would include cascading failures of other critical infrastructures. As such, the 7-day onsite storage of diesel fuel will be inadequate, and the 30-day diesel fuel resupply requirement would most likely be difficult or perhaps even impossible to execute when disruptions of transportation, communication, and even refining

342 International Journal of Nuclear Security; Sherrell R. Greene Nuclear Power: Black Sky Liability or Black Sky Asset? https://trace.tennessee.edu/ijns/vol2/iss3/3/

343 Nuclear Regulatory Commission, Regulatory Guide 1.137 Fuel Oil Systems for Emergency Power Supplies, June 2013, https://www.nrc.gov/docs/ML1230/ML12300A122.pdf

of diesel fuel are hindered. For the U.S. NPP fleet, other options beyond safe shutdown would likely prove to be much safer.

Before looking into Small Modular Reactors (SMRs), which potentially offer the prospects of blackstart capability and inherently safer designs, it is important to consider the possibility that the current U.S. fleet of LWRs could be modified to meet blackstart requirements. Additionally, the legacy nuclear power plants could be used more effectively in flexible power modes as has been demonstrated both in the U.S. and abroad.

B. Flexible Operations as an Enabler for Extended Safe Shutdown and/or Load Following

A hypothetical example of a technically feasible, "safe mode" that would utilize flexible operations at a LWR plant could be as follows: A 2000 MWe NPP has four reactors, each with a maximum Rated Thermal Power (RTP) output of 500 MWe. In the event of a LOOP event, the present "safe shutdown" SOP is stopping all 4 reactors and running diesel generators to provide the estimated 10% or 200 MWe required to run cooling pumps and control equipment. Instead, tripping 3 of the reactors, and running the 4th at 40% RTP, would provide the same 200 MWe power necessary to keep the plant in this new, modified safe shutdown mode for months to years with no need for refueling. The time would be extended if each of the reactors was put into the role of powering house loads sequentially.

Flexible operations that allow deviations from 100% RTP are not just useful for safer shutdown, but can also allow load following, a key requirement for blackstart and grid restoration. While the above example is hypothetical, operating NPPs at levels below 100% RTD is not.

France leads the world in having the largest share of its electricity – 63% - supplied by nuclear power.[344] The French utility Électricité de France (EDF)

344 Statistica; Share of Nuclear Power in Total Domestic Electricity Generation in 2022, by Select Country, https://www.statista.com/statistics/270367/share-of-nuclear-power-in-the-power-supply-of-selected-countries/

employs flexible operations, including rapid load following — from 100% RTP down to 30% RTP – with minimal reactor trips, while maintaining stable power at these various power levels.[345] The EDF example demonstrates that it is possible for legacy nuclear plants to operate more flexibly.[346],[347]

Such flexible operation would have important safety and fuel security advantages. It could also allow legacy LWRs in the U.S. to play an important, constructive role as either blackstart or next start units, enhancing the Black Sky resilience of the overall BES.

- **Recommendation:** It is important to have the existing nuclear power plant fleet looking at how they could become Black Sky assets to the BES. U.S. government resources should be applied to look at the licensing and technical barriers preventing the existing nuclear fleet from becoming black sky assets through more flexible operation.

C. Small Modular Reactors (SMRs)

The U.S. DOE is interested in the resilience benefits offered by a proposed new breed of Advanced Small Modular Nuclear Reactors (SMRs).[348] SMRs, though they are currently still in design phase, could be designed to operate and startup without an external power source and therefore be cable to support grid recovery. DOE highlights five key resilience features that SMRs could be designed to provide: blackstart, islanding, underground construction, fuel

345 POWER; Flexible Operation of Nuclear Plants Ramps Up, https://www.powermag.com/flexible-operation-of-nuclear-power-plants-ramps-up/

346 Sherrell R. Greene (2018) "The Key Attributes, Functional Requirements, and Design Features of Resilient Nuclear Power Plants (rNPPs)" Nuclear Technology, 204:2, 131-146 https://doi.org/10.1080/00295450.2018.1480213

347 A more thorough analysis of the ability of the U.S. LWR reactor fleet to operate with a similar level of flexibility is beyond the scope of this analysis. As of the time of this writing, however, the Electric Power Research Institute (EPRI) is conducting research on the ability of U.S. nuclear plants to be prepared, and possibly leveraged, for Black Sky events.

348 DOE, Office of Nuclear Energy; 5 Key Resilient Features of Small Modular Reactors https://www.energy.gov/ne/articles/5-key-resilient-features-small-modular-reactors

security, and modularity.[349]

Blackstart: SMRs could be designed to start themselves without outside grid power, the essential definition of a blackstart generator. Ther ability not only to (re)start themselves, but to also operate very flexibly – the ability to operate at varied levels of power output, not just at or very near 100% of rated thermal power (RTP), as discussed above – means SMRs could also be designed to be more capable of "load following" meaning they can play a constructive role in BES blackstart, next start and grid stabilization operations.

Islanding: To be useful as either a blackstart or next-start generator along the blackstart cranking path, a generator must be able to switch rapidly from a small electrical island (where it starts itself) to larger islands (next-start) and then larger grid sections as blackstart builds to a stable recovery.

Underground Construction: DOE notes that underground construction makes SMRs less vulnerable to extreme weather events, earthquakes, electromagnetic pulses and other intentional destructive acts. While this could be true for certain scenarios, EIS would caution that it should not be taken as a blanket statement that underground construction always reduces vulnerability to Black Sky events.

> The need for a reliable and standalone self-starting power generator in the face of a Black Sky event is clear and with the emergence of SMRs it is imperative that blackstart capability be a part of their design.

Fuel Security: The ability to generate power for months to years without refueling is a tremendous component of Black Sky resilience and blackstart in particular. Such fuel security would be second only to hydropower and geothermal power plants. The fuel security aspect makes SMRs a very attractive option for Black Sky resilience, when fuel resupply will almost certainly be extremely challenging to impossible.

Modularity: Many advantages include minimal components, especially electrical components. Construction would not be bespoke and done on sight as was/is true of the traditional large LWRs. Modularity offers the advantage

349 Ibid

of standardized parts and designs, and factory assembly. The new designs are also intentionally inherently safer, using passive cooling / shutdown features rather than active measures that require power, undamaged safety systems, and operator intervention. These safety advantages translate directly into Black Sky resilience as well. Systems that function properly and safely without grid power are the definition of Black Sky resilient.

- **Recommendation:** The need for a reliable and standalone self-starting power generator in the face of a Black Sky event is clear and with the emergence of SMRs it is imperative that blackstart capability be a part of their design.

Note: All SMR designs discussed below are still in the design phase, and ensuring blackstart capability in their final design and construction is essential.

D. Current, Promising Small Modular and Advanced Reactor Designs[350], [351]

1. Small Modular Light Water Reactors

Small modular light water reactors are smaller LWRs that currently make up the bulk of the U.S. Nuclear Power Plant (NPP) fleet. "Small" is defined by the National Academies of Science (NAS) as less than 300 MWe. The smaller size can enable inherent "passive" safety features that do not require active systems to maintain cooling. Instead, internal pressure and gravity are harnessed, eliminating a crucial safety failure mode associated with failure of cooling pumps to operate, both during normal operation, and – importantly – during blackouts or other LOOP events.

350 National Academies of Science; Laying the Foundation for New and Advanced Nuclear Reactors in the United States (2023), https://www.nationalacademies.org/our-work/laying-the-foundation-for-new-and-advanced-nuclear-reactors-in-the-united-states

351 Sherrell R. Greene (2020) "How Nuclear Power Can Transform Electric Grid and Critical Infrastructure Resilience", Journal of Critical Infrastructure Policy, Volume 1, Number 2, Fall/Winter 2020, pp 37-72; https://www.jcip1.org/greene.html

In addition to the safety advantages, operators familiar with traditional LWRs should be able to operate SMRs with less additional training than other designs. The NAS reports that the "most mature" small modular LWR is the NuScale Power Module (NPM). The Nuclear Regulatory Commission approved NPM for Design Certification in 2022, and a further design is under consideration. This could allow such plants to be built and operated within the next 10 years. Unfortunately, the planned installation of the first NPM was recently canceled.[352]

Important Note: The NPM is currently the only SMR design that has the NRC Design Certification.

Figure 7.8: the NuScale Small Modular Light Water Reactor[353]

The NPM is in the Pressurized Water Reactor (PWR) category. PWRs operate using two water circuits, primary and secondary. The primary coolant water circuit is used to keep the reactor core at the proper operating temperature and to transfer heat from the reactor core to the secondary circuit. The primary coolant water is kept at high pressure, which prevents the primary coolant water from boiling and undergoing a phase transition to steam. The secondary

352 Science; Adrian Cho, Deal to Build Pint-Sized Nuclear Reactors Canceled, https://www.science.org/content/article/deal-build-pint-size-nuclear-reactors-canceled

353 DOE, Office of Nuclear Energy; Advanced Small Modular Reactors (SMRs), https://www.energy.gov/ne/advanced-small-modular-reactors-smrs

coolant water circuit does allow the water to boil and change to steam, which is used to drive the turbine to generate electricity.

This description of primary and secondary cooling loops is generic to all PWRs regardless of their size. Three more important, unique features of the NPM are: (1) natural circulation cooling in the primary loop, (2) integral reactor vessel with the primary heat exchanger inside the reactor vessel, (3) submersion of the reactor vessel in a very large pool of water that, under accident conditions, provides a passive heat sink and water source to keep the reactor core cooled. A schematic of a PWR is shown in Figure 7.9.

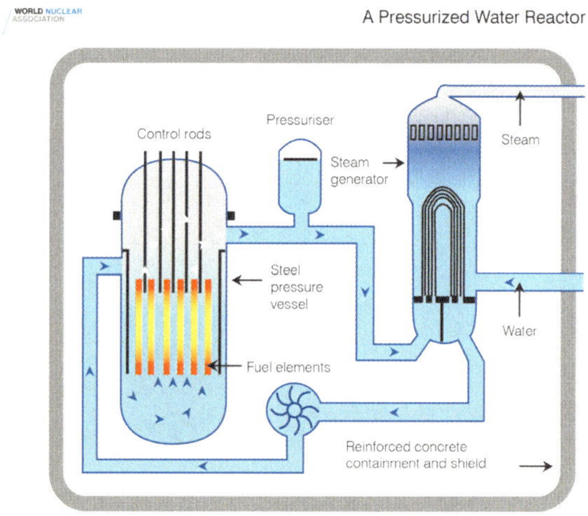

Figure 7.9, Pressurized Water Reactor[354]

The U.S. Department of Energy is also interested in advanced, small modular, boiling water reactor (BWR) designs as well. A BWR only uses one primary cooling water circuit. The water that cools the core is allowed to convert to steam, which directly drives the electricity-generating turbine. A schematic of the BWR design is shown in figure 7.10.

354 World Nuclear Association Image Library, Pressurized Water Reactor, https://www.world-nuclear.org/gallery/reactor-diagrams/pressurized-water-reactor.aspx

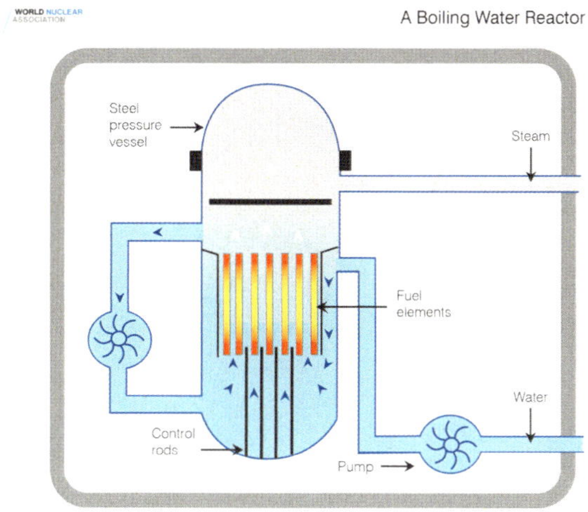

Figure 7.10, Boiling Water Reactor Design[355]

One such SMR under development is the Hitachi BWRX-300, with an output power of 300 MWe.[356] This design also incorporates passive cooling, allowing the reactor to maintain safe operation with optimal reactor pressure and temperature for up to seven days without external power or operator intervention in the event of a LOOP event. Importantly, The BWRX-300, shown below in Figure 7.11, can be operated flexibly and has the ability to load follow, requisite for blackstart operations.

355 World Nuclear Association Image Library, Boiling Water Reactor https://www.world-nuclear.org/gallery/reactor-diagrams/boiling-water-reactor.aspx

356 DOE, Office of Nuclear Energy; First U.S. Small Modular Boiling Water Reactor Under Development, https://www.energy.gov/ne/articles/first-us-small-modular-boiling-water-reactor-under-development

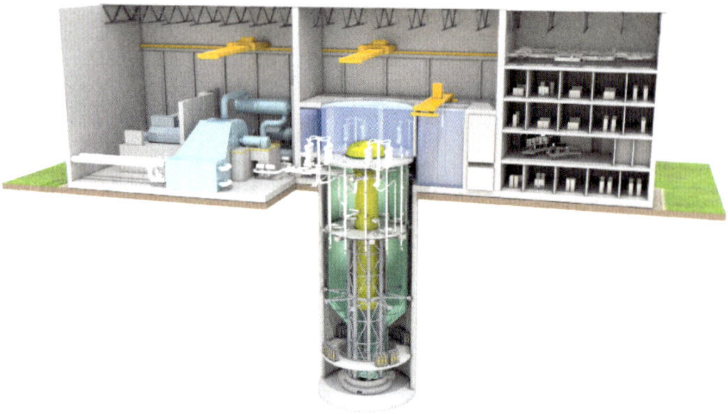

Figure 7.11, the Hitachi BWRX-300 Small Modular Reactor[357]

2. Liquid Metal-Cooled Fast Reactors

Liquid metal-cooled fast reactors are also considered by the NAS to be a "relatively mature design concept". The leading design in the U.S. is the Sodium Fast Reactor (SFR), and uses liquid sodium, rather than water, as the coolant. The SFR design is also characterized by inherent safety features and use of passive systems. There are a few design types, but the current leading type is the "pool-type" design, which uses a pool of sodium to cool the reactor core. These units can differ in size from 100-1500 MWe. A schematic of an SAF is shown in Figure 7.12.

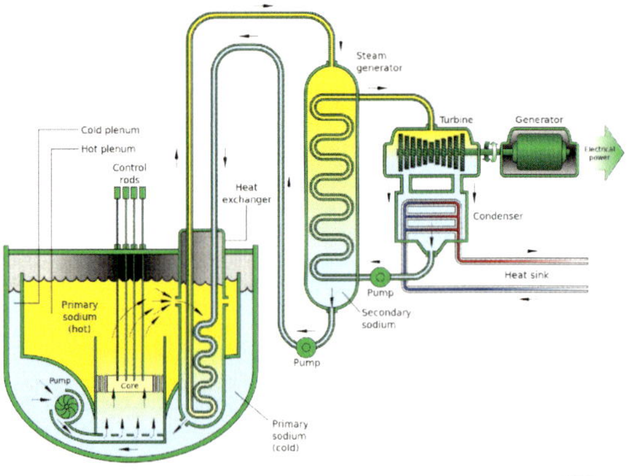

Figure 7.12, Pool-type Sodium-cooled Fast Reactor[358]

357 Ibid

358 Wikipedia, Sodium-cooled Fast Reactor, https://en.wikipedia.org/wiki/Sodium-cooled_fast_reactor

3. High Temperature Gas-Cooled Reactor

The high temperature gas-cooled reactor (HTGR) employs inert gas (typically carbon dioxide or helium) to cool the core and as the heat transfer medium. The uranium fuel is encapsulated in small graphite blocks or balls (called "pebbles") rather than fuel rods. While reactor power levels and system architectures vary, HGTRs typically offer inherent safety by virtue of their low power densities, use of graphite-based fuels and moderators, and passive cooling using natural heat conduction, thermal radiation, and thermally-driven gas circulation. No active cooling/heat-transfer systems are used.

The systems can be configured to operate as base load or more flexible load-following modes. A demonstration project of one such HTGR, X-energy's Xe-100, is planned to be demonstrated in the next 10 years. The plant construction time is claimed to be only four years, and the modular components are to be factory made, transported by road, and assembled on-site. The design has four 80 MWe reactors for a total of 320 MWe. The pebbled fuel design enables online refueling, spent fuel retrieval and storage, rather than requiring core shutdown to accomplish these processes. A schematic of the HTGR is shown in Figure 7.13.

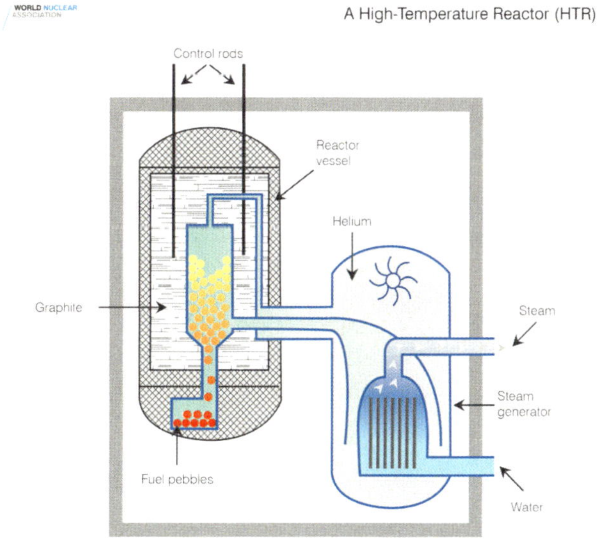

Figure 7.13, High-Temperature Gas-Cooled Reactor[359]

359 World Nuclear Association Image Library, High-Temperature Reactor https://www.world-nuclear.org/

4. Gas-Cooled Fast Reactor

The gas-cooled fast reactor is a related design to the Sodium Fast Reactor but uses helium as the coolant/heat transfer medium. As of this writing, there are no proposed/anticipated demonstration projects.

5. Fluoride Salt-Cooled High-Temperature Reactor

The fluoride salt-cooled high temperature reactor (FHR) utilizes lithium fluoride and beryllium fluoride rather than liquid sodium as in the SFR, While multiple concepts exist, the most mature is Kairos Power's pebble bed concept. Kairos recently obtained NRC approval for construction of its "Hermes" 35 MW demonstration reactor at a site in Oak Ridge, TN .[360]

6. Molten Salt Reactor

The molten salt reactor (MSR) design is unique in that the uranium fuel is dissolved in the molten salt coolant. An experimental reactor of this type was built and successfully operated at Oak Ridge National Lab during the 1960s. Two companies (TerraPower in the U.S.) and Terrestrial Energy (in Canada) are pursuing development of MSR concepts and are engaged in regulatory discussions with their respective nuclear safety regulators.

7. Micro-Scale Nuclear Reactors

These very small reactors produce less than 10 MWe. They are mostly envisioned for non-BES, microgrid applications that require self-contained, resilient, dedicated power, including industrial sites, military bases, or small off-grid communities. Some designs are adapted from space-based designs, while others are smaller versions of liquid metal-cooled and gas-cooled reactors.

These microreactors can be fully built at a fabrication facility and transported, installed, and operated at the use site, with some minimal supporting

gallery/reactor-diagrams/high-temperature-reactor.aspx

360 Oak Ridge National Laboratory; Flouride-Salt-Cooled High-Temperature Reactors, https://www.ornl.gov/content/fluoride-salt-cooled-high-temperature-reactors

infrastructure. They are intended to generate power autonomously, or with minimal on-site operation staff. Regulatory challenges currently exist, however, for transportation of such fully fueled reactors.

The U.S. Air force has selected one such reactor, the Oklo Microreactor, to provide electricity at Eielson Air Force base in Alaska.[361] Project Pele at the Department of Defense's Strategic Capabilities Office is also interested in mobile microreactors.[362]

E. Blackstart Capability of Small Modular Reactors

Small Modular Reactors can, and should, be designed to have the ability to self-start autonomously, without the need for outside power from the bulk power system. Their smaller size means correspondingly less power is required to start them, accomplished by even smaller on-site power sources. This capability, along with their ability to operate flexibly allows them to follow load and therefore be a much more resilient asset as blackstart and next start units.[363]

SMRs would also have another inherent capability to act as blackstart units, especially in a Black Sky environment: the inherent fuel security of having up to ten years of fuel onsite. This, combined with the inherent safety features and passive cooling systems, would also eliminate another Black Sky vulnerability: running out of fuel.

361 EnergyTech; Rod Walton, U.S. Air Force Selects Oklo Microreactor for Power Project at Alaskan Base, https://www.energytech.com/distributed-energy/article/21274573/us-air-force-selects-oklo-microreactor-for-power-project-at-alaskan-base?utm_source=ET+Transition&utm_medium=email&utm_campaign=CPS230928179&o_eid=4708I6058334B1I&rdx.ident[pull]=omeda|4708I6058334B1I&oly_enc_id=4708I6058334B1I

362 DOD Strategic Capabilities Office, Project Pele, Mobile Nuclear Reactor; https://www.cto.mil/pele_eis/

363 Sherrell R. Greene (2020) How Nuclear Power Can Transform Electric Grid and Critical Infrastructure Resilience, Journal of Critical Infrastructure Policy, Volume 1, Number 2, Fall/Winter 2020, pp 37-72; https://www.jcip1.org/greene.html

- **Recommendation:** The current U.S. LWR fleet should consider SMRs as sources of emergency backup power. Should the main plant trip offline due to a LOOP event, and therefore rely on on-site backup power to run the active cooling, control, and safety systems of the bulk of the nuclear power plant, the SMR could provide reliable emergency power to keep the LWRs in safe shutdown mode for as long as any conceivable Black Sky event could last. In addition, incorporating SMRs as emergency backup power-capable units at current nuclear power plants offers the advantage that the owners and operators of the plant already have the day-to-day experience and expertise of safely and effectively operating nuclear power plants.

V | FUEL CELLS AS BLACKSTART ASSETS

While diesel generators are currently the most used asset for blackstart operations, the push towards cleaner and more sustainable alternatives make fuel cells more attractive.

Diesel generators are not designed for long-term use, and they require service, replacement parts and fuel resupply. Fuel cells, by contrast, are designed for long term or even constant/continuous use. They have no moving parts and therefore require very limited servicing, often limited to simple annual air-filter replacement. Fuel cells also offer longer run-times than similarly sized diesel generators[364], an important factor in Black Sky emergency backup power or blackstart applications.

364 Altergy; Alternative Backup Power: Comparing Batteries and Diesel Generators to Hydrogen Fuel Cells, www.altergy.com/wp-content/uploads/2019/03/Altergy_CleanEnergyBackupPwr_WP_Telco.pdf

Diesel generators also have other limitations and challenges. They produce a constant current level, so aren't able to follow load.[365] Diesel fuel degrades over time and requires upkeep, monitoring, and sometimes replacement. A fine sediment and/or gum can form as diesel fuel oxidizes over time. Aerobic and anaerobic bacteria can also grow in diesel fuel tanks. These unwanted contaminants can clog fuel filters, so frequent filter changes are required to keep the engine operational.[366]

> Diesel generators are not designed for long-term use, and they require service, replacement parts and fuel resupply. Fuel cells, by contrast, are designed for long term or even constant/continuous use.

Fuel cells are described in greater detail in chapter VIII. They are quickly becoming an attractive alternative to diesel generators for emergency backup power generation. Fuel cells contain all the characteristics for blackstart, though currently their limitation for most blackstart operations is in their small size. They can be chained together to add up to larger power, like batteries in series, but their primary utility at this stage of development is in emergency backup power rather than BES blackstart. Many are large enough, though, for distribution grid black start (DGBS), discussed above.

Utility companies have already begun installing megawatt-size fuel cells at power stations. The Dominion Bridgeport fuel cell park in Connecticut houses a 14.9 MW array, capable of blackstart,[367] and two 30 MW cell arrays are installed at Delmarva Power substations in Delaware.[368] As this trend continues, fuel cells are expected to provide BES-level power applications, including blackstart.

Microgrids are being funded by several states in the northeastern U.S. to ensure reliable power during storms or other outages. Connecticut is a prime

365 Ibid

366 Ibid

367 DOE-AMO Workshop; Pinakin Patel, Fuel Cells for Grid Modernization, Opportunities and Needs, www.energy.gov/sites/default/files/2016/02/f29/Patel%20AMO%20Workshop%20Austin%20presentation%20FCE%202-10-16.pdf

368 Plant Engineering; Fuel Cells Find New Applications, https://www.plantengineering.com/articles/fuel-cells-find-new-applications/

example with several microgrids that include fuel cells, providing power for emergency shelters, first responders, and public works facilities. After Hurricane Sandy hit CT very hard in 2012, the state was determined to be ready for the next such storm, and support the deployment of resilient power technologies.

The Connecticut Microgrid Program was established to provide a foundation for reliable energy supply to critical infrastructure. Figure 7.14 below lists some of the microgrids established through the program, showing the effectiveness of fuel cells.[369] The University of Bridgeport has a Megawatt-scale fuel cell microgrid that supplies "80 percent of the power for critical facilities at the University of Bridgeport and cuts the energy costs by 20 percent during normal operations."[370] Beyond the financial savings, "more than 7000 tons of carbon-dioxide, 64 tons of sulphur oxides, and 28 tons of nitrogen oxides" are not emitted when using fuel cells instead of diesel generators.[371]

> Batteries in series are also capable of grid black start. The Tesla Mega Pack at Plus Power in Oahu, Hawaii is now designated as that island's black start asset.
>
> For Tesla, it is quite a large project with 565 MWh of Megapacks to deploy, but the company has deployed larger systems, like the recently commissioned Moss Landing Project, which has a capacity of 730 MWh and is planned for an expansion to over 1GWh.

- **Recommendation:** Fuel cells can offer blackstart capability, the only current limitation is their size. As fuel cell arrays become larger, they should be considered a blackstart asset, first at microgrid and distribution levels, and eventually for BES applications.

  Note: Batteries in series are also capable of grid black start. The Tesla Mega Pack at Plus Power in Oahu, Hawaii is now designated as that

369 Resilient Power; Connecticut Department of Energy and Environment Microgrids Program, https://www.cesa.org/wp-content/uploads/CT-DEEP-case-study-Nov-2014v2.pdf

370 Microgrid Knowledge; University Achieves 20 Percent Savings with Smart Financing and a Fuel Cell Microgrid, https://www.microgridknowledge.com/google-news-feed/article/11431442/university-achieves-20-percent-savings-with-smart-financing-and-a-fuel-cell-microgrid

371 Ibid

island's black start asset. "For Tesla, it is quite a large project with 565 MWh of Megapacks to deploy, but the company has deployed larger systems, like the recently commissioned Moss Landing Project, which has a capacity of 730 MWh and is planned for an expansion to over 1GWh."[372]

CT DEEP AWARD WINNERS			
ROUND 1			
Project	**Facilities**	**Generation**	**Grant Amount**
UConn Depot Campus/Storrs	Campus buildings	400 kW fuel cell, 6.6 kW PV	$2,144,234
City of Bridgeport-City Hall/Bridgeport	City hall, police station, senior center	(3) 600 kW natural gas microtrubines	$2,975,000
Wesleyan/Middletown	Campus, athletic center (public shelter)	(1) 2.4 MW and (1) 676 kW natural gas CHP reciprocating engine	$693,819
University of Hartford-St. Francis/Hartford	Dorms, campus center, operation building	1.9 MW diesel (existing), 250 kW diesel, 150 kW diesel	$2,270,333
SUBASE/Groton	Various buildings and piers	5 MW cogen turbine, 1.5 MW diesel	$3,000,000
Town of Windham/Windham	2 schools (various public purposes)	130 kW natural gas, 250 kW solar, 200 kWh battery; (2) kW diesel	$639,950
Town of Woodbridge/Woodbridge	Police stations, fire station, Dept. of Public Works, Town Hall, high school, library	1.6 MW natural gas, 400 kW fuel cell	$3,000,000
City of Hartford-Parkville Cluster/Hartford	School, senior center, library, supermarket, gas station	600 kW natural gas	$2,063,000
Town of Fairfield-Public Safety/Fairfield	Police station, emergency operations center, cell tower, fire HQ, shelter	50 kW natural gas recip. engine, 250 kW natural gas recip. engine, 47 kW PV	$1,167,659
ROUND 2			
City of Milford	Parsons complex, middle school, senior center, senior apartments, city hall	(2) 148kW natural gas CHP units, 120KW PV, 100kW battery storage	$2,909,341
University of Bridgeport	Dining hall, rec center, student center, 2 res buildings as shelters, police station	1.4 MW fuel cell	$2,180,898

Figure 7.14, Connecticut Microgrid Systems.[373]

372 https://electrek.co/2022/06/27/tesla-megapacks-replace-hawaii-last-remaining-coal-plant/

373 https://www.cesa.org/wp-content/uploads/CT-DEEP-case-study-Nov-2014v2.pdf

8

CHAPTER EIGHT

MEETING RESILIENCE REQUIREMENTS AGAINST BLACK SKY HAZARDS

I | PLANNING

A. Minimum Grid Concept

Resilience planning for the BES first requires prioritization and ranking of assets or components of the system from 'absolutely critical' down to 'non-essential'.

'Absolutely critical' systems or assets include: blackstart generator; next start generator; fuel for both blackstart and next start generators; the complete electric cranking path between them (lines, substations (or even/often specific transformers and busses within the substation) and critical loads (which could include additional [after-next start] generation); protection and control (P&C) throughout the path; communications; and synchronization (some version – synchrophasors or digital fault recorder with synchrophasor data option, or synchroscopes).

As electric utilities become faced with an increasing reliance on the use of electricity in modern society, the expectation of increased reliability and resiliency becomes a driving factor. This expectation is exacerbated by increasing threats such as electromagnetic pulse (EMP) events or widespread cyberattacks. Utilities are therefore faced with four primary blackout scenarios:

Blackout Types			
Type1	No or minimal damage	2003 Northeast blackout	Cascading blackout
Type2	Widespread apparent physical damage	Hurricane Maria	Natural disaster
Type3	Invisible physical damage	Starfish Prime	High-altitude EMP burst
Type4	Invisible cyber damage	Ukraine Power Grid Cyber Attack	Cyber attack

Blackstart plans address Type 1 blackouts while other System Restoration plans are inhibited by discovery, response and triage. In type 2 blackouts the damage is readily apparent by the equipment damage and typically impacts a subset of the transmission / distribution systems. For Type 3 and Type 4 blackouts the damage to equipment may not be apparent. Recovery may take several months due to the limited number of specialized technicians investigating the extent of the damage, challenges for those technicians to operate in a Black Sky environment, and limited availability of spare equipment.

Cyber or EMP events would likely occur with little or no warning, and traditional mutual assistance personnel – linemen, contractors, and electricians – will not be the workforce required to initially respond and assess damage. Investigating, repairing, or replacing components that may be damaged during a cyber or EMP event will require significant time for trained technicians.

In an effort to speed the recovery for Type 3 and Type 4 blackouts, utilities are working to harden their systems for causes of potential damage. The number of facilities that will need to be hardened becomes an overwhelming task due to the sheer volume of equipment that may be impacted. An average utility may have thousands of relays and control systems to protect in addition to computer-controlled hardware that operate electrical equipment and generating plants. For EMP events, power transformers, capacitor bank designs, in addition to control systems, also must be considered.

B. Strategies for Load Pickup and Prioritization, Especially for Societal Continuity

In power restoration operations, where blackstart is the most challenging scenario, there currently exist a couple of schools of thought:

Prevailing paradigm: restore the most meters in the shortest amount of time. This typically involves restoring power to houses/apartments and buildings. This practice is effective and acceptable when power outages are short (minutes to hours to 3 days, max) but is the wrong policy for Black Sky events, which by definition extend to weeks or months.

Black Sky, Minimum Grid thinking: prioritize critical loads that provide the most societal "life support" in/to the grid (gas pipelines and other fuels, nuclear plants); public safety (law enforcement, government, volunteer organizations active in disasters (VOADS), public health (water, wastewater), economic stability (financial networks), homeland/national security (military readiness, critical infrastructure protection/security), public information (media).

A second, important consideration, is the load size and electrical resilience of the load, for grid stabilization. Large resistive loads are best for grid restoration/ stability. Inductive loads like directly connected motors/pumps that require huge "inrush" currents and stabilizing reactive power are less optimal from a blackstart standpoint, but in many scenarios must be prioritized to be energized because of their societal resilience value. Large data centers are on the other end – they are essentially resistive loads (maybe with the exception of the air conditioners/cooling) so those loads are "good" from a blackstart perspective, but the load itself is sensitive, meaning that several stops and starts (due to grid/ island instability) could be damaged by voltage/current surges.

C. Minimum Grid Planning Steps[374]

The purpose of the Minimum Grid approach is to build upon the traditional restoration plans by using some of the blackstart cranking paths that have been developed for a Type 1 cascading blackout and Type 2 event restoration plans where much of the grid is damaged due to storms and cannot quickly return to service. This approach ensures critical loads are restored rapidly by providing a defined action plan for personnel.

The Minimum Grid is a subset of the grid that the utility manages and will serve critical distribution buses by selecting a subset of the transmission system by using developed blackstart cranking paths, blackstart units and next tier generation units that can energize these critical buses.

1. Minimum Grid Selection

a. Average Load Assumption

Assume an average load period, not a peak load, in the service area. Peak periods are infrequent. However, if an outage requiring blackstart were to occur during such a period, rotating blackouts could be implemented until the peak period concludes.

b. Distribution Bus Selection

Define the distribution buses that serve the most load – do not consider redundancy. Verify that critical, first priority loads are included within the selected distribution system.

c. Transmission System Selection

Define the transmission system required to serve the selected distribution buses utilizing existing blackstart cranking paths wherever possible.

374 Minimum Grid Planning Steps were included in EPRO IV – EMP, restated here for ease of reference.

2. Configure a Blackstart Plan for the Selected Minimum Grid

a. Develop a blackstart plan, based on two key considerations:

i. The selected Minimum Grid transmission system; and

ii. Critical, first priority loads. These would include, for example, nuclear plants, electric utility command and control facilities, government security installations, state and local government command and control facilities, water/wastewater plants, and other high priority infrastructure facilities.

Note: Given the public fear that will likely result from these types of events, it is important to include command and control facilities for restoration of service as soon as possible to coordinate and aid recovery.

b. Minimum Grid Blackstart Plan Simulation

Run the selected Minimum Grid Blackstart Plan system through a power flow simulation to make sure voltage during restart falls into acceptable levels, and that relay protection settings are adequate to protect the system.

c. Facility Hardening

Once the blackstart plan is validated through simulation, develop a plan to secure and verify the included substations, relay systems, transmission sites, generating stations, and other components against EMP, cyberattacks, and other Black Sky hazards. This should include ensuring that all transmission sites have synchrophasors or digital fault recorders with a synchrophasor data option, so they are visible to system operations. If possible, make sure synchroscopes are installed, or digital synchroscopes are available for these stations, to assist with restart.

d. Training, Spares and Emergency Supplies

Include training and drilling for the Minimum Grid Blackstart Plan in the regular training plan for all personnel who will be required to respond to implement the Plan – relay technicians, IT personnel, electricians, etc. Drills should include: how to investigate damage from the event, where to initially respond, how to understand when they are to respond, where to get replacement units, emergency supplies, etc.

Key Takeaways

1. Energizing the Minimum Grid is (Only) Step One

   This Minimum Grid is not redundant and does not include additional assets to account for equipment unavailability. Once it is energized, personnel must work to build redundancy into the network. This would include strategic placement of distributed energy resources along this path.

2. Minimum Grid Planning Provides a Basis for Prioritized EMP Hardening

   Defining the Minimum Grid provides a framework for selecting and prioritizing critical sites for EMP hardening and where EMP protected communications will be needed for restart.

3. 3. Minimum Grid Planning Provides a Basis for Establishing Minimum Prioritized Spares and Emergency Supplies

By defining the requirements for assessing and repairing the sites included in the Minimum Grid, utilities can make an initial estimate for critical spares, tools, digital protective equipment, and other components that may be difficult to acquire in the disrupted societal scenarios Black Sky events will create.

II | ASSET HARDENING

Asset hardening is not strictly limited to electric grid assets, it is also necessary to include critical load hardening if the load is within the blackstart cranking path. If the critical load is damaged or destroyed by the threat/event and cannot accept electrical power, it will not be available to participate in the cranking path, and therefore slows or stalls blackstart. In addition, that critical load's function will not be available even if the grid is quickly restored. The techniques, practices, and technologies for asset hardening are essentially the same, whether hardening assets within the electrical grid/system or critical loads.

Hardening is often threat/event specific. Sparing, redundancy, and strategic islanding discussed in the next section, are more universal. Asset hardening options for the Black Sky Threats/Hazards are:

A. Cyber Protection

All entities in the U.S. BES are required to comply with the North American Electric Reliability Corporation (NERC) Critical infrastructure Protection (CIP) standards which include both cybersecurity and physical security requirements. There are currently 13 CIP Standards[375], twelve for Cybersecurity and one for Physical Security:

CIP-002-5.1a BES Cyber System Categorization — Identify and categorize BES cyber assets for the application of cyber security requirements commensurate with the adverse impact that loss, compromise, or misuse of those BES cyber assets could have on the reliable operation of the BES.

CIP-003-8 Security Management Controls — Specify consistent and sustainable security management controls that establish responsibility and accountability to protect BES cyber systems against compromise.

CIP-004-6 Personnel & Training — Minimize the risk against compromise that could lead to misoperation or instability in the BES from individuals accessing BES cyber systems by requiring an appropriate level of personnel risk assessment, training, and security awareness.

CIP-005-7 Electronic Security Perimeter(s) — Manage electronic access to BES Cyber Systems by specifying a controlled Electronic Security Perimeter in support of protecting BES Cyber Systems against compromise that could lead to misoperation or instability in the BES.

CIP-006-6 Physical Security of BES Cyber Systems — Manage physical access to BES cyber systems by specifying a physical security plan in support of protecting BES cyber systems against compromise.

CIP-007-6 System Security Management — Manage system security by specifying select technical, operational, and procedural requirements in support of protecting BES cyber systems against compromise.

CIP-008-6 Incident Reporting and Response Planning— Mitigate the risk to the reliable operation of the BES by specifying incident response

375 Industrial Defender; What is NERC CIP: The Ultimate Guide, https://www.industrialdefender.com/blog/what-is-nerc-cip

requirements, including the reporting of attempted or actual compromises of entities' systems to national cybersecurity entities.

CIP-009-6 Recovery Plans for BES Cyber Systems— Recover reliability functions performed by BES cyber systems by specifying recovery plan requirements.

CIP-010-4 Configuration Change Management and Vulnerability Assessments — Prevent and detect unauthorized changes to BES cyber systems by specifying configuration change management and vulnerability assessment requirements in support of protecting BES cyber systems from compromise.

CIP-011-2 Information Protection — Prevent unauthorized access to BES cyber systems from compromise that would affect the stability of the BES.

CIP-012-1 Communications Between Control Centers — Maintain situational awareness and reliable bulk electric system operations by protecting the confidentiality and integrity of Real-time Assessment and Real-time monitoring data transmitted between Control Centers.[376]

CIP-013-2 Supply Chain Risk Management — Mitigate cyber security risks to the reliable operation of the Bulk Electric System (BES) by implementing security controls for supply chain risk management of BES Cyber Systems.

CIP-014-3 Physical Security — Identify and protect transmission stations and transmission substations, and their associated primary control centers, that if rendered inoperable or damaged as a result of a physical attack could result in instability, uncontrolled separation, or cascading within an interconnection.

Recommended cyber security/hygiene practices include:

- Password management, other access requirements, 2 Factor Authentication (2FA).
- Isolation of business enterprise (IT) network from OT network (air gap);
- Operational/Organizational separation of IT & OT;
- Bespoke (if possible), minimum operating systems for OT;

376 https://www.federalregister.gov/documents/2020/02/07/2020-01595/critical-infrastructure-protection-reliability-standard-cip-012-1-cyber-security-communications#:~:text=NERC%20states%20that%20Reliability%20Standard,data%20while%20being%20transmitted%20between

- Rigorous 2FA required for remote access to OT;
- Backup Data (onsite, and offsite), backup Operating System Copies, etc. (Redundancy solution)

B. Earthquake Protection for Natural Gas Pipelines

The National Earthquake Hazards Reduction Program (NEHRP)[377] is a partnership between FEMA, NIST, USGS, and NSF. Two key factors for earthquake hardening are understanding the earthquake risk at any particular location, and designing, building, or in some cases, retrofitting assets to withstand earthquakes.

For the continental U.S., the earthquake hazard maps provided by NEHRP are below in Figure 8.1.

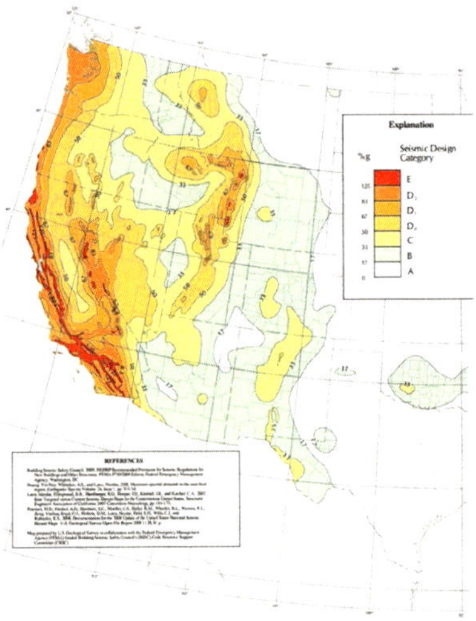

Figure 8.1.a. Earthquake risk map for the Western half of the U.S.

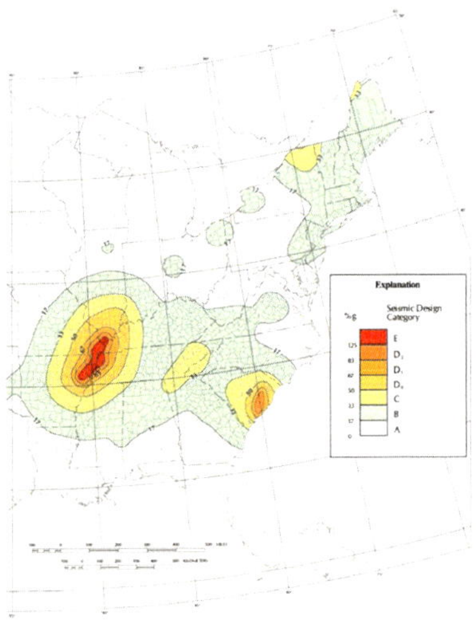

Figure 8.1.b. Earthquake risk map for the Eastern half of the U.S.

377 National Earthquake Hazards Reduction Program; https://www.nehrp.gov/

From a Black Sky blackstart and next start perspective, we must look beyond the immediate damage to assets in the earthquake zone and to the larger disruption to the BES and supporting infrastructures, in particular the U.S. natural gas pipeline system, shown in Figure 8.2. below.[378]

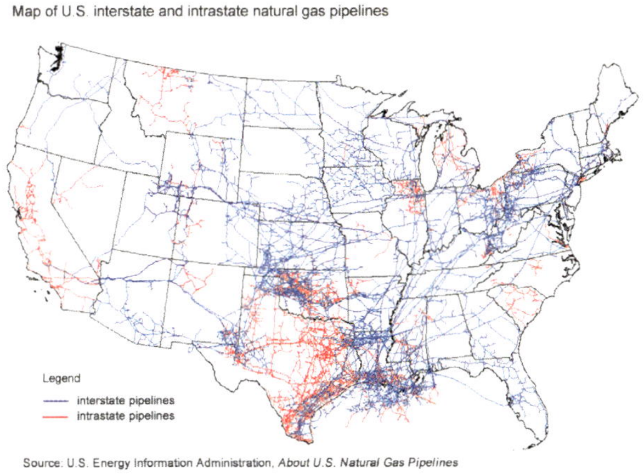

Figure 8.2. Major natural gas pipelines in the continental United States.

Viewing Figure 8.2. and Figure 8.1.b, we see that a large number of interstate natural gas pipelines, running from the U.S. Gulf Coast to the U.S. Northeast and Midwest run through the New Madrid Earthquake zone, the large seismic zone in Figure 8.1.b. Disruption or destruction of these pipelines will starve the Northeast and Midwest of natural gas, currently the principal fuel for blackstart and especially next-start generators (see Chapter V). While the concern for natural gas pipeline disruption also exists on the U.S. West Coast, it is especially acute for those that run though the New Madrid Earthquake Zone, largely due to a large disparity in earthquake risk awareness between the West Coast and the Central U.S.

- **Recommendation:** Natural gas-fed blackstart and next start units in the U.S. should invoke a 2-layer strategy in protecting their fuel sources. First, a secondary/backup fuel type such as fuel oil must be available on site to use at those generators in the event of natural gas pipeline

378 U.S. Energy Information Administration; Natural Gas Explained: Natural Gas Pipelines, https://www.eia.gov/energyexplained/natural-gas/natural-gas-pipelines.php

disruption. Second, public and private sector investment in protecting natural gas pipelines is needed. The most common example, shown in Figure 8.3 below, shows the Alyeska pipeline in Alaska, which uses flexible joints mounted on rails to protect the pipeline from Earthquake Damage.[379]

Figure 8.3, Alyeska Pipeline. Flexible joints mounted on rails allow the pipeline to move without breaking during earthquakes.

C. Electromagnetic Pulse (EMP) and Geomagnetic Disturbances (GMD) Protection

The EIS Council has done extensive work on EMP protection, with a special focus on the electric power sector. Preceding this Blackstart-focused EPRO Handbook, the entirety of the EPRO IV handbook "Electromagnetic Pulse Protection Best Practices"[380] is dedicated to this subject, with a special focus on High-Altitude EMP (HEMP). In addition, EPRO 1, Chapter 2[381] discusses GMD protection options extensively. For ease of reference, however, below is a short summary of the EMP/GMD risks and recommended hardening practices.

379 Alyeska Pipeline; EARTHQUAKES, https://www.alyeska-pipe.com/pipeline_facts/earthquakes/
380 EIS Council, EPRO Handbook IV - EMP, https://eiscouncil.org/black-sky/
381 EIS Council EPRO Handbook I, https://eiscouncil.org/black-sky/

HEMP is created when a nuclear weapon is detonated 30 miles or more above the earth. High energy gamma rays released by the explosion interact with the atmosphere, generating radio waves at frequencies between 1 GHz down to near 0 Hz (quasi-DC) and lasting up to 4 minutes. The pulse is usually described as consisting of 3 phases (See Figure 8.4).

E1 lasts for only a microsecond, with Frequencies between 1 GHz and 10 kHz, and a maximum electric field strength of 50 kV/m. E1 poses a threat to electronic devices such as relays, digital control systems, computers, and communications equipment.

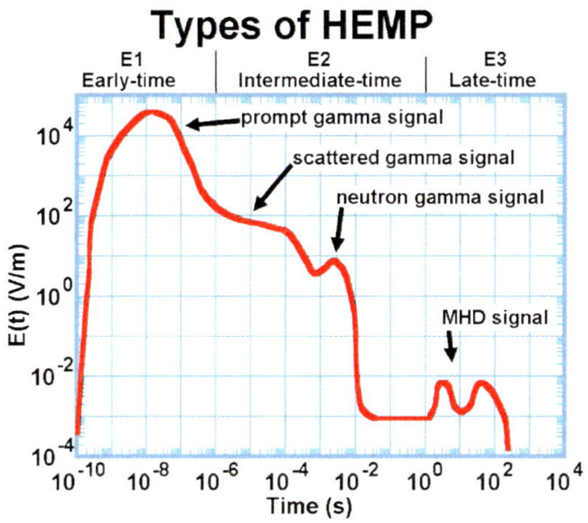

E2 follows E1, lasting for one hundredth of a second, with a maximum Electric field strength of 100 V/m. E2 has an electric field profile similar to lightning and is of less concern due to the widely-deployed use of lightning protection within the electric sector.

The final phase, E3, is driven by disturbance of the Earth's magnetic field. This results in a low-frequency (quasi-DC) electric field with maximum field strengths of 50 V/km, lasting up to 5 minutes. E3 will not harm electronics but can cause damage to large transformers within the electric grid. Geomagnetic Disturbances (GMD) caused by solar Coronal Mass Ejections (CMEs) striking the Earth's magnetic field are very similar to E3 but can last for hours to days.

1. E1 Hardening

It is important to remember that while the EMP Pulse is created by a nuclear weapon, it is fundamentally a radio wave pulse. The radio waves interact with metal conductors to cause unwanted currents. When these radio waves impinge directly on an object like a digital relay, this is termed the "radiative" threat. When they impinge on a metal cable that is attached to the relay, like a power or ethernet cable, this is termed the "conductive" threat. Research highlighted in EPRO IV shows that the conductive threat is more likely to cause equipment disruption or damage, but both should be addressed.

E1 Protection is accomplished by proper shielding, grounding, and surge protection techniques and devices. Faraday shielding is accomplished by surrounding the protected object with a conductive (metal) shield, which is properly grounded. Cables can also be grounded and shielded.

Devices can also be protected from conductive threats by using surge protection devices such as metal oxide varistors (MOVs) at the cable/device portal. Use of fiberoptic instead of metal cables is also effective, as fiberoptic cables do not carry electric currents. The use of wireless communications is also effective, as WiFi and Bluetooth operate above 2.4 GHz and so are out of band of E1.

- **Recommendation:** Designs to protect critical electronic equipment from E1 threats should be phased in for critical minimum grid assets along blackstart cranking paths. NERC should reconstitute the EMP Working Group, and promulgate requirements for E1 protection of assets, especially blackstart cranking path assets.

2. E3 Hardening

Unlike E1, E3 does not harm microelectronics. The principal at-risk assets for E3 are large, High Voltage (HV) or Extra High Voltage (EHV) transformers operating on the BES. The quasi-direct currents (DC) that flow in the BES due to E3 cause a "bias" in transformers, which are designed for alternating current (AC). This causes stray "eddy" currents that heat structural elements of the transformer, leading to damage. The "bias" also induces higher harmonics above the fundamental 60 Hz frequency that can be damaging to loads or induce tripping of transformers.

NERC has adopted requirements to protect the BES from GMD[382]. In brief, they require transmission operators in the North American grid to monitor geomagnetically induced currents (GIC) in their systems; to analyze their

382 NERC; TPL-007 Transmission System Planned Performance for Geomagnetic Disturbance Events, www.nerc.com/pa/Stand/Reliability%20Standards/tpl-007-4.PDF

transmission system to determine maximum GIC flows against a NERC-supplied benchmark GMD event; and to develop a mitigation plan, which can be a combination of hardware and operational posturing of the system.

NERC also had a working group on EMP (now paused) but has not yet adopted requirements for it. Even though the physics are similar, an important difference between EMP and GMD is that with GMD, a coronal mass ejection can be seen on the sun, which gives between 15 hours to several days warning that it is approaching, and when the CME is 1 million miles from Earth, the DSCOVR satellite can measure the magnetic field strength and orientation, providing 15-60 minutes of warning with this additional information. An EMP would likely have no warning.

The TPL-007 Requirements should be expanded to include relevant EMP E3 considerations, such as higher peak fields but shorter time durations, and without latitude scaling (since an EMP burst could be anywhere over the U.S., whereas GMD is typically stronger at higher latitudes).

- **Recommendation:** The TPL-007 Requirements should be expanded to include relevant EMP E3 considerations, such as higher peak fields but shorter time durations, and without latitude scaling (since an EMP burst could be anywhere over the U.S., whereas GMD is typically stronger at higher latitudes).

  Operational approaches to allowing the BES to ride though a GMD include keeping more lines and transformers in service (to essentially "spread out" the GIC to make it less intense), or the opposite – removal of weak transformers from service so they are not exposed to GIC. Other approaches are to use transformer designs with higher GIC withstand capabilities (use of non-magnetic steel structural elements, dual winding or three winding transformers versus autotransformers[383], core-type versus shell type); use transformers with larger capacity margins (de-rating); and to dampen or block the GICs by using

383 Autotransformers can be designed and built to better handle GIC, but this is not presently the case for the vast majority of autotransformers.

series capacitors on transmission lines, and capacitors or resistors on transformer neutral grounding wires. Table 8.1 below, taken from the first EPRO Handbook,[384] lists various transformer protection options and evaluation of effectiveness.

Transformer Hardening Options	Option Evaluation Criteria						
	Confidence Level	Cost Effectiveness	Transfomer GIC Decrease	Transformer Harmonic Decrease	System Availability and Reliability	Overall Grid GIC Decrease	Hardening/Procedural
Transfomer Type	+	+	0	0	+	0	H
Transformer GIC Witstand Specification	+	+	0	0	+	0	H
Transformer De-rating	+	-	0	0	0	0	H
Additional Transformers	+	-	+	+	+	0	H
Spare Transformers	+	-	N/A	N/A	N/A	N/A	P
Series Capacitors	+	0	+	+	+	+	H
Pre-planned Tranformer Disconnection	-	-	N/A	N/A	-	0	P
Pre-planned Transformer Tripping - At Threshold	+	+	+	+	-	0	H/P
- On E1 Sensor	+	+	+	+	N/A	0	H/P
Neutral Ground Blocking	+	+	+ (- for auto transformers)	+ (- for auto transformers)	+	+	H
Neutral Ground Reducing	+	+	+ (- for auto transformers)	+ (- for auto transformers)	+	+	H
Do Nothing / System As Is	-	-	-	-	-	0	N/A

Key for the Evaluation Matrix
The following nomenclature is used to indicate a rough assessment of performance of the different protection options for each evaluation criterion:
"+" – Option performs well on criterion
"0" – Option performs adequately on criterion
"-" – Option performs poorly on criterion
For the last column, "H" indicates a hardware hardening approach and "P" indicates a procedural approach

Table 8.1, Transformer Protection and Evaluation Matrix

384 EIS Council; EPRO Handbook I, https://eiscouncil.org/black-sky/

D. Extreme Weather (Wind, Rain, Snow, Ice, Fire, Flood, Extreme Temperatures) Mitigation [385]

The most commonly experienced threats to electric grid assets are naturally occurring extreme weather events. Such events typically cause geographically limited Grey Sky events, but extended or multiple, successive events could cause enough grid disruption to become a full Black Sky event.

Recommendation: The most vulnerable asset along black start cranking paths are the transmission and distribution power lines. Robust vegetation management (tree trimming) along the cranking path is required to prevent the most common type of line break and will also prevent fire damage by removing the fire source. NERC has a vegetation

385 Note: figure sources: https://kogt.com/transmission-trouble-hurting-progress/ and https://ie.unc.edu/2021/06/06/piehler-presents-preliminary-findings-to-nc-senate-select-committee-on-flood-resilience/

management standard which applies to all BES entities[386], but extra care and margin along cranking paths is recommended. On sub-transmission (below 200kV) and distribution systems, the vegetation management standards, if any, are maintained by state and local Public Utility Commissions, and vary depending on location. These standards should also be met and exceeded wherever possible to ensure the physical connectivity of the cranking path to critical loads. Physical strengthening of support poles and towers themselves is also recommended for the cranking path lines.

Recommendation: Generators and substations along the cranking path should be housed in extreme wind/weather-proof shelters. Such shelters on the blackstart cranking path should provide protection against many or most of the hazards listed above and below: built to also withstand extreme earthquakes, provide Faraday Shielding, and kinetic attack. Flood mitigation techniques such as flood walls or elevated assets can also be used. Emergency backup generators and their fuel reserves, battery arrays, etc. must be similarly protected.

E. Kinetic Attack (Rifle fire, Explosives) Protection

The NERC CIP Standards, while mostly focused on cybersecurity, do include one standard on Physical Security.[387] In 2022, nearly 1,700 physical attack incidents were reported to the Electricity Information Sharing and Analysis Center (E-ISAC).[388] The NERC CIP-14 Standard applies only to critical assets at all transmission substations operating at 500 kV or above,

386 NERC; Transmission Vegetation Management NERC Standard FAC-003-2 Technical Reference, www.nerc.com/pa/Stand/Project%20200707%20Transmission%20Vegetation%20Management/FAC-003-2_White_Paper_2009Sept9.pdf

387 NERC; CIP-014-3 – Physical Security, www.nerc.com/pa/Stand/Reliability%20Standards/CIP-014-3.pdf

388 TRC; Fuller and Fitzgerald, NERC Files Report on Effectiveness of CIP-014 Physical Security Standard, https://www.trccompanies.com/insights/nerc-files-report-on-effectiveness-on-cip-014-physical-security-standard/#:~:text=CIP%2D014%20was%20conceived%20to,and%20all%20500%20kV%20substations.

and some substations between 200 – 500 kV, depending on weighting criteria that quantifies the connectivity of the substation. It does not apply to generation substations.[389] The standard is not prescriptive but relies on companies covered by the requirements to take a risk-based approach to physical security.

Recommendation: the scope of the substations covered by CIP 14 – Physical Security should be expanded to include all generation and transmission assets along identified blackstart cranking paths.

389 NERC; CIP-014-3 – Physical Security, www.nerc.com/pa/Stand/Reliability%20Standards/CIP-014-3.pdf

III | SPARING / REDUNDANCY / STRATEGIC ISLANDING

While asset hardening is often threat-specific, redundancy / sparing / stockpiling / defense-in-depth are essentially threat-agnostic best practices. They are highly effective almost without exception. The principle (and probably only) downside/challenge is that they are often viewed as "expensive/inefficient/wasteful" from within the "Blue Sky" paradigm.

Spare assets must be employed quickly and easily to be useful in a Black Sky (or really, any) emergency. Having a "spare" HV transformer, for example, that must be moved from one site to another, using special heavy equipment, is a process that can take days to weeks, even on a Blue Sky day. In a highly disrupted Black Sky environment, such "spares" are probably stranded, and not able to be used. Effective sparing requires assets that can be quickly swapped in to replace the primary asset if it is damaged or otherwise non-functional.

Sparing best practices for Black Sky blackstart must account for asset protection and security of the spares. If spares are stored in vulnerable configurations that would allow them to be damaged or destroyed by the same event that compromised the primary asset, the sparing effort is ineffective and unnecessarily costly. Worse still, in-service assets which are counted as spares will likely not be available post-event. This will happen either because they are themselves damaged or destroyed by the same Black Sky event, or they require disconnection and reassignment to take the place of the primary asset. This requires technical and personnel intervention, as well as possibly difficult/challenging transportation of the asset. A mixture of spare types and storage locations should be deployed.

Examples of shortcomings include use of co-located, connected SCADA or other computer assets. If a backup SCADA is under power, closely co-located and connected to the system during an event, it could likely suffer the same damage as the primary system it is intended to replace. Backup/redundant systems must truly be isolated and protected from Black Sky events if the expectation is that they will be rapidly and effectively substituted for the damaged primary asset. A resilient backup SCADA system must be only intermittently connected to the primary system to allow capture of relevant operational data but must spend the majority of its time in an (preferably) unpowered and system-isolated state. At the same time, it must be stored in a location that will allow it to be employed rapidly in a Black Sky event.

Beyond SCADA systems, minimum grid blackstart cranking path assets include:

- UPS, battery, or other ancillary power sources;
- Switchgear
- Relays
- Synchroscopes
- Transformers
- Backup communications systems

- **Recommendation:** At the “minimum grid” level, the recommended standard should be 2N for all critical blackstart cranking path assets. 2N reliability means each critical asset has one dedicated backup/spare. This is necessary as any minimum grid asset is, by definition, essential to form and maintain the cranking path. One damaged, out-of-service asset in the "minimum grid" cranking path eliminates that path. Therefore, redundancies of all minimum grid assets are required to make the cranking path possible.

A further consideration at the BES level, beyond the minimum grid, that would have immediate resilience benefits is to move from N-1 to N-2. If N-1 is the "Reliability Standard", then N-2 = (N-1)-1 could be defined as the Reliability Plus Resilience Standard.

A qualitatively distinct category of resilience strategy aims to reduce the probability that failure will propagate from one asset to another along the power grid's topology following a primary or initiating failure event. The concept is to keep the Black Sky from spreading from one part of the grid to another, limiting total consequences. For example, a generating station could be disabled by a Black Sky threat, but it is possible through protection strategies and transmission grid engineering to limit the chances of that failure taking down other transmission assets via a cascading failure event. Strategically placed islanding and microgrids along with optimized transmission topology are important "all hazard" resilience strategies that are robust against a wide range of threats.[390], [391]

390 Mouco, A., et al. Resilience of Power Grids to Catastrophic Cascading Failures; Proc. CIGRE U.S. National Committee 2022 Grid of the Future Symposium. 2022.

391 Mouco, A., et al. A Discussion of Resilience in Power Grids; Proc. CIGRE U.S. National Committee 2023 Grid of the Future Symposium, 2023.

IV | PERSONNEL

A. Training

For an organization to be Black Sky blackstart ready, a culture of resilience must be developed. This must be driven by leadership and involve at minimum all key operational, planning, and regulatory personnel. Organization and personnel readiness for this extreme contingency requires pre-event resilience investment, and detailed, executable, and flexible event response plans.

Planning will not translate into effective blackstart execution unless personnel are given the opportunity to practice/train for the Black Sky event, including multiple contingencies. Perhaps the most important characteristic of true blackstart plans and training is the assumption that blackstart will need to occur in the framework of a highly disrupted, and possibly/likely a contested, environment. A common shortcoming of current blackstart planning assumptions and requirements is that any blackout is limited to the BES and is focused on voltage collapse, with minimal or no permanent BES asset damage assumed. Further, it is often assumed that other critical infrastructures like communications, transportation, fuel supply and resupply, finance, water/ wastewater, etc., will be functioning normally.

- **Recommendation:** Black Sky blackstart plans and training must include these "reasonable worst-case" contingencies. Further, that training should ensure that operators have the ability to "self-deploy" and act autonomously because they have detailed, pre-event playbooks with them.

Black Sky blackstart plans and training must include these "reasonable worst-case" contingencies. Further, that training should ensure that operators have the ability to "self-deploy" and act autonomously because they have detailed, pre-event playbooks with them.

B. Redundancy and Availability Measures

Redundancy will, in essentially all cases, ensure availability. In the current state of affairs, black start generators are not redundant, and in fact, are "oversubscribed" in many places within the BES. At the BES level, redundant blackstart cranking paths, and the assets within them, are required to help ensure availability of blackstart and next start power. Adequate fuel sources must be available, and protected, along the blackstart cranking path. Diversity of blackstart and next start generation sources including dual-fuel capability, and on-site or in-cranking-path energy storage are necessary.

Personnel redundancy is also a requirement. In a Black Sky environment, some personnel will likely be unavailable to participate in operations. If knowledge of blackstart operations is limited to a small number of expertly trained operators, it becomes more likely that in a Black Sky blackstart event, some of that already small number of expert operators will not be available (transportation challenges, workers or their family members may be hurt, vulnerable, or otherwise restricted, etc.) Cross training of electric system personnel in blackstart operations (described above) is also essential.

V | ESSENTIAL, WIDELY DEPLOYED BLACK SKY-CLASS RESILIENCE CAPABILITIES

A. Emergency Communications

In every NERC GRID-EX, a key finding has always been that current communications will fail.[392] Another closely related finding is that blackstart is impossible without some basic means of communication. The more basic – e.g., voice only communications on handhelds with only direct line-of-sight broadcast and reception – the more difficult blackstart will be. In a basic communication scenario, black start is not impossible, but will require a labor-

392 E-ISAC; Public GridEx Reports I – VII; https://www.eisac.com/s/gridex

intensive operation wherein grid operators must relay verbal commands from the RTO to TOs, who in turn must relay commands back and forth to personnel at generators, substations, and critical loads.

Those generator operators must adjust blackstart and next start power outputs, as well as synchronize frequency, sometimes "by hand." Operators at substations must open and close breakers to energize buses that connect power to critical loads in sub-transmission or distribution systems.

Communication of data is also extremely important. While it is possible to blackstart "by hand" using voice communications on radios and manual synchroscopes, auto-synchrophasors are much preferred to provide the power, reactive power, and frequency matching needed for grid island stability and prevention of subsequent grid re-collapse. Passing this data directly, rather than translating and relaying using voice, vastly increases the likelihood of success during blackstart operations.

Another key advantage to using data is that leveraging data allows operators to restore at scale, whereas voice is slow and cumbersome. Another reason data is necessary is for decision support for Chaos Management, discussed in the next section.

> In every NERC GRID-EX, a key finding has always been that current communications will fail. Another closely related finding is that blackstart is impossible without some basic means of communication. The more basic – e.g., voice only communications on handhelds with only direct line-of-sight broadcast and reception – the more difficult blackstart will be.

Obviously, it is also essential that communications equipment can function without grid power, so self-powered systems are a prerequisite for blackstart. In addition, it is a best practice that the communication system be owned and operated by the electric system companies/operators themselves, for two reasons: a) it is highly likely that the standard telecoms systems will be disrupted, either by the precipitating event (storm, cyberattack, etc.) itself, or as a cascading failure due to lack of grid power; and b) in such a highly chaotic environment, more people and entities will be trying to communicate, overtaxing the bandwidth of the system, even if it is not damaged and has enough backup power to operate.

Telecom networks are key strategic assets just like the grid. In a worst-case scenario, the government and other entities will need these resources because it is also what they rely on to communicate. This often pushes critical infrastructure down on the telecom priority list during emergencies. Government Emergency Telecommunications Service (GETS) and Wireless Priority Service (WPS) all ride commercial telecom, as do First Net and Guard Net, etc. This is where the telecoms will be forced to focus during emergencies. Finally, communications are also a part of the National Infrastructure Advisory Council (NIAC) where the focus is primarily on strategic capabilities that may not be public (e.g. government and military communications in the event of a nuclear war.)

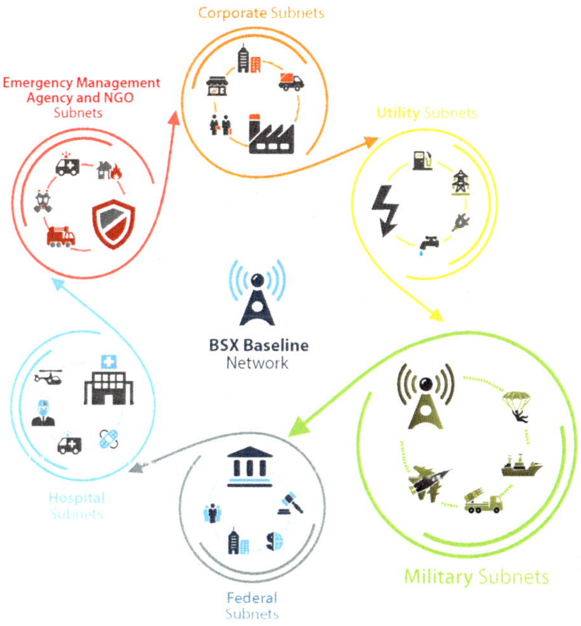

Figure 8.5.a: High level overview of the BSX® System, an interoperable, all hazard-protected emergency communication and coordination system which will operate during emergencies and catastrophes without requiring connectivity to the nation's traditional telecommunications networks.

EIS Council, in cooperation with industry partners, has created and piloted the Black Sky Emergency Communications and Coordination System – BSX® – that meets these criteria (See Figure 8.5 below). It is self-powered, can communicate using multiple, redundant pathways (radio, optical fiber), can self-form mesh networks, "harvest" frequencies that are not in use because the normal user of that frequency spectrum is down, and "relinquish" frequencies when the normal user returns to operation, and is hardened against all Black Sky threats.

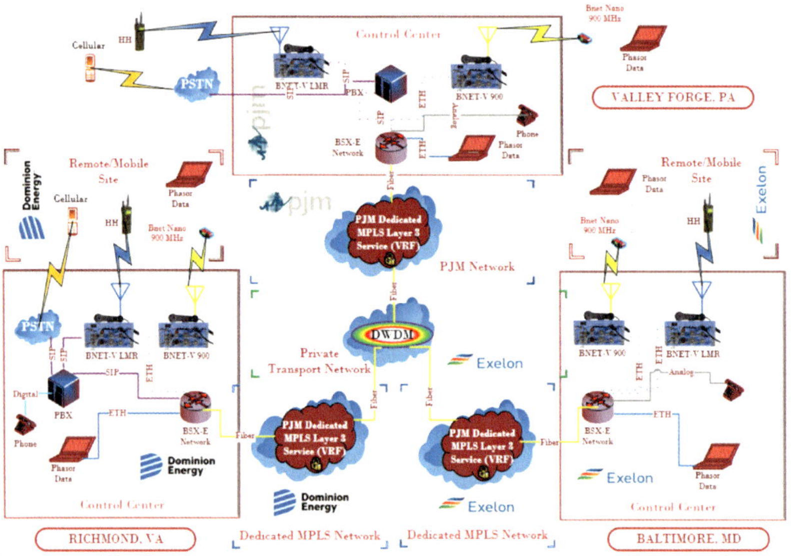

Figure 8.5.b: The BSX® Pilot 2 Project, currently connecting Dominion, Exelon, and PJM

You can find a more in-depth discussion on communicating in a black sky environment in the Electric Infrastructure Protection (EPRO®) Handbook III Cross-Sector Coordination and Communications in Black Sky Events published by the EIS Council.[393]

Encouraged by the successful demonstration of BSX Pilot 2, the power industry has decided to take the next step and is moving forward with the ESWeb™, the first one-hundred percent utility owned and operated multi-company, multi-state high-bandwidth resilient communications network. You can find out more about the ESWeb™ by visiting the EIS Council website at: https://eiscouncil.org/esw/

393 EIS Council, EPRO Handbook III, https://eiscouncil.org/black-sky/

B. Chaos Management

Our modern, highly integrated infrastructure system-of-systems, or metagrid, has become incredibly efficient, effective, and convenient. However, it has also become complex—in virtually all senses of that word—and therefore Black Sky power outages can be both caused by and lead to disruptions across all infrastructure sectors. Perhaps worse, these disruptions can cascade through and across infrastructure sectors in patterns that are difficult to predict in advance, let alone contain as they unfold. The likelihood of unwanted, unintended consequences in response, recovery, and restoration operations is nearly 100% without the aid of sophisticated situational awareness and decision support tools for decision makers, operators, and responders that takes full account of this complexity and all its consequences.

The ability to effectively game-out, plan, and train for complex catastrophes requires interactive, digital twin models of infrastructure functions and their interdependencies with sufficient fidelity to permit analysis of multiple disaster scenarios, as well as multiple response scenarios, pre-event. Such models would also support decision makers during disasters, both as readily interpretable yet information-dense windows into the developing situation and as just-in-time forecasting tools that can play out possible consequences of multiple proposed response actions virtually before action is taken on the physical infrastructure system.

The Global Infrastructure Network Optimization Model (GINOM®) has been under development by EIS Council for six years for just this purpose. GINOM® is a large-scale, multi-sector, geographically aware critical infrastructure simulation operating system based on techniques from agent-based modeling and simulation. As such, GINOM® is a unique, disruptive chaos management platform.

As a critical feature, GINOM® functions as a digital twin of the facilities of multiple interdependent infrastructure organizations and supporting systems. Scale is achieved via functional integration of an essentially unlimited number of organizations' existing data feeds, operating systems, and simulations. With this structure, EIS Council does not need to digitally duplicate the vast number of constantly changing infrastructure systems necessary to run a reasonable fidelity simulated environment, a daunting task that would require an army of dedicated software developers working around the clock to stay abreast of and digitally duplicate every change. Instead, GINOM® can integrate the models and simulations already being produced by infrastructure operators into a usable whole. By relying exclusively on top-line functional information, there is no need for proprietary or sensitive information sharing in its operation. Individual organization's simulations can run on their own secure, audited hardware under their own security protocols while still participating in a large-scale GINOM® simulation.

An additional feature that sets GINOM® apart from other approaches to the chaos management problem is that its agent-based methodology does not rely solely on projecting historical data into the future. It also takes account of the physical and logical relationships and interdependencies among different infrastructure facilities and their dynamic behavior. Taking account of this network structure in addition to historical data allows GINOM® to identify the class of Black Sky events that emerge from the network structure, a near-impossible task for Artificial Intelligence or Machine Learning based methods. It is impossible to train an Artificial Intelligence or Machine Learning model to predict massive, cascading multi-infrastructure failures that have never happened before, because there is simply no data to use to train such models. GINOM® has no trouble simulating such phenomena and could even be used to generate synthetic datasets that could be used by downstream efforts.

Figure 8.6: Screenshot of multi-infrastructure interface from current GINOM® prototype. A time snapshot of a global cyber and EMP crisis simulation..

C. Long Duration Power Cells

Emergency Backup power is important in many locations, including hospitals, emergency services, water/wastewater, nuclear power plants, industry, finance, communications, governance, etc.

While this handbook is focused on ensuring that blackstart capabilities are robust and can quickly and effectively re-start the grid, it is also prudent for critical infrastructure facilities to ensure that they have long-duration backup power that can sustain basic operations of the facility for up to a month, should it take that long for grid power to be restored. In such a long-duration power outage, other critical infrastructures will also be disrupted.

In such an environment, it is important for continuity of operations plans to recognize that fuel resupply for emergency backup generators will most likely be unavailable, as transportation in particular will be a massive challenge. In addition, backup diesel generators are not typically designed to run continuously for a month or more, thus putting critical facilities at risk during extended duration emergency situations. Other options for long-term emergency backup power such as solar-plus-battery storage or fuel cells, are needed. While grid-level storage is still a challenge, facility level storage, adequate for at least one month of minimum Black Sky continuity of operations, is possible. Critical facilities must have the ability, on site, to maintain minimal function in an extended outage, supported by facility-level, long duration power capability.

1. Fuel Cells for Long Duration Emergency Backup Power

Fuel cells are an alternative type of power generation with high energy density, ideal when large energy quantities are required. This makes fuel cells a very viable option for sustainable power plants. Able to supply green emergency power to data centers and hospitals, fuel cells have begun replacing diesel generators within microgrids.

Fuel cells work by generating electric current through a reaction of Hydrogen and Oxygen. The fuel cell has an anode and a cathode, similar to a battery. Electrons are split from hydrogen at the anode. The electrons flow through a circuit, while the hydrogen nuclei (protons) flow across the cell to combine with oxygen and the electrons that have traveled through the circuit to the cathode. The result is electric current, and the only emissions are water vapor.

This simple process is shown in Figure 8.7 below.

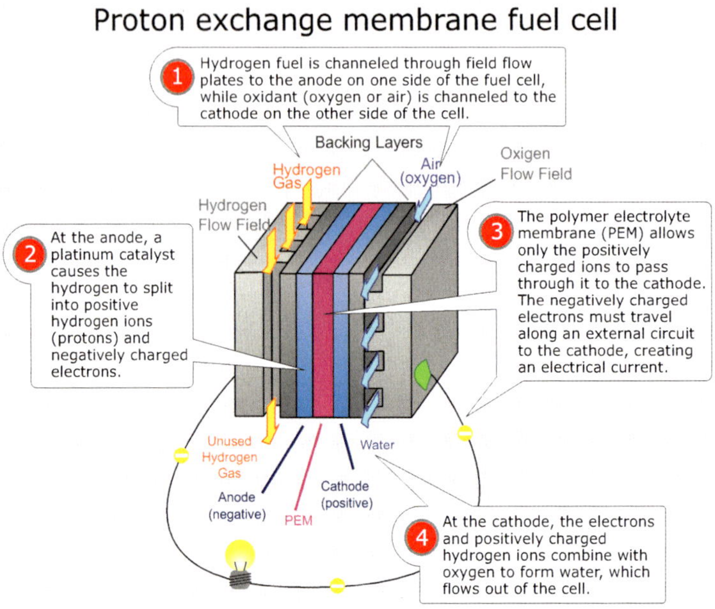

Figure 8.7: How Fuel Cells Work[394]

At the 2021 UN COP 26 climate conference in Scotland, Rolls-Royce Power Systems unveiled its new "future-proof mtu fuel cell system for C02-free power generation"[395]. The fuel cells they presented are "scalable and integrated complete solutions for reliable power generation in the megawatt range and are to be available as standard systems from 2025"[396]. Each fuel cell module has a net power output of 150kW, and modules can be stacked to output in the megawatt range. For example, 1.5MW could be achieved by stacking 10 of these modules.

394 DOE, Office of Energy Efficiency and Renewable Energy, How Fuel Cells Work, https://www.fueleconomy.gov/feg/fcv_pem.shtml

395 MTU; Fuel Cells and the Quest for Green Power, https://www.mtu-solutions.com/na/en/stories/technology/research-development/fuel-cells-and-the-quest-for-green-power.html

396 Ibid

GenCell[397] is a company that produces an array of modular fuel cells that can be both integrated into the grid and operate off grid. GenCell has three main types of cells: BOX, REX, and FOX. BOX is a long-duration backup power cell designed specifically to support telecommunications infrastructure. REX is geared toward backup power for utilities and would be able to provide sustainable power to substations during an outage at 5kW per unit. FOX is able to provide off-grid power to areas disconnected from the main grid during an outage. GenCells could not only replace diesel generators as a more sustainable black and next start asset, but would solidify communications, supplement substations, and bolster grid severances between the three cell options.

Fuel cells utilize hydrogen, one of the most energy dense non-radioactive fuels. The battery is more efficient and able to output more power than the fuel cell, however, batteries fall short when it comes to weight constraints. One fuel cell system with a tank containing 3kg of hydrogen has an energy density of around 120 MJ/kg, while a 625kg Li-ion battery has an energy density of only 0.567 MJ/kg.

For example, fuel cells are a viable option for semi-trucks because it allows them to "refuel as fast as conventional diesel rigs and also weigh less"[398]. Electric trucks driving across the country will have to stop for hours of recharging along the way, whereas a hydrogen fuel cell truck with zero-emissions is able to refuel as fast as the conventional gas truck.

For batteries, 90% of power reaches the power train while 10% is often lost in transmission. Fuel cells, though, lose around 40% in transmission and retain only 60%. The U.S. Department of Energy projected solar-to-hydrogen conversion efficiency is in the range of 30–40% for 2025[399]. Hydrogen electrolyzers range in quality and size, while the power source of the electrolyzer is a factor in its efficiency, with advanced technology electrolyzers achieving much higher power to hydrogen conversion efficiencies as high as 90%.

397 GenCell; https://www.gencellenergy.com/

398 New York Times; Truck Makers Face a Tech Dilemma: Batteries or Hydrogen?, https://www.nytimes.com/2022/04/11/business/electric-hydrogen-trucks.html

399 ScienceDirect; Solar-to-Hydrogen Conversion Efficiency, https://www.sciencedirect.com/topics/engineering/solar-to-hydrogen-conversion-efficiency#:~:text=Various%20PV%2DE%20systems%20have,%25%20%5B7%2C8%5D.

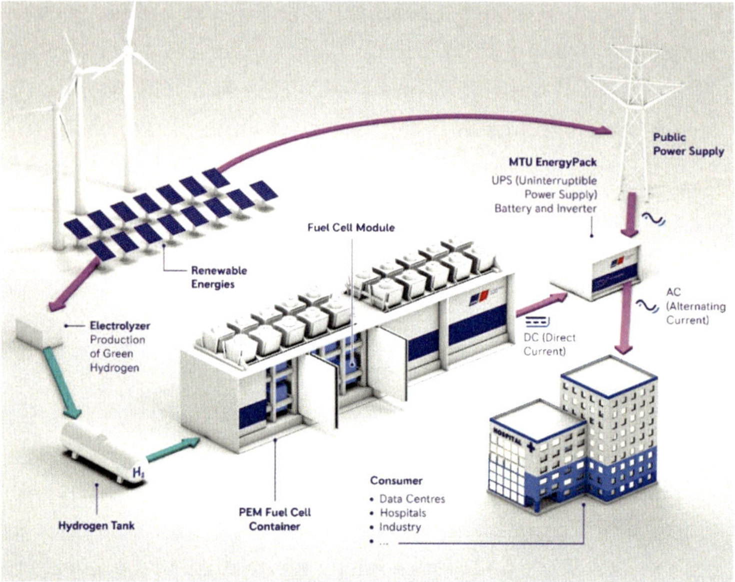

Figure 8.8, Renewable Fuel Cell System[400]

According to the U.S. Department of Energy, fuel cells are an "attractive options for reliable backup power.[401] When Superstorm Sandy hit the Caribbean and the East Coast, "fuel cells provided emergency backup power to at least 100 telecommunications towers in both the Bahamas and the Northeast United States"[402]. In the wake of Hurricane Irene in 2011, "ReliOn fuel cells kicked on at 56 Sprint cell towers, and Doosan fuel cells maintained power at both a storm shelter at South Windsor High School and a Whole Foods location in Connecticut"[403].

Backup power from fuel cells is a viable option for not only communications, but many other parts of our critical infrastructure. Grocery stores rely on power to maintain perishables, putting them at risk in the event of an extended outage. Once a black sky event occurs, the flow of goods is disrupted, and it becomes

400 MTU; Fuel Cells and the Quest for Green Power, https://www.mtu-solutions.com/na/en/stories/technology/research-development/fuel-cells-and-the-quest-for-green-power.html

401 DOE, Office of Energy Efficiency and Renewable Energy, Fuel Cells Powering Critical Infrastructure in Disasters, https://www.energy.gov/eere/fuelcells/articles/fuel-cells-powering-critical-infrastructure-disasters-fact-sheet

402 Ibid

403 Ibid

much harder to get supplies like food. Ensuring that food already in affected areas can maintain refrigeration so it doesn't spoil will make it easier to keep those areas supplied until power is restored. "Several supermarket chains, including Whole Foods, Safeway, Stop & Shop, Albertson's, Star Market, and Price Chopper, have invested in fuel cell backup systems at some of their new stores"[404], hoping to mitigate the impact of losing power. During Hurricane Irene, the first fuel cell ever installed in a grocery store, a Whole Foods in Connecticut, successfully retained power through the storm with its 200kW fuel cell.[405]

Hospitals are a critical part of our core infrastructure that requires constant power, where losing power means losing lives. Power is needed to maintain life support for patients, to conduct operations, and to refrigerate perishable medical supplies. Fuel cells are a reliable solution for hospitals to maintain power during outages and care for current and future patients. "During Hurricane Sandy, several hospitals were forced to evacuate patients due to malfunctioning backup generators". In response to the issue, "St. Francis Hospital in Hartford, Connecticut, has installed two 400kW fuel cells for stationary power"[406].

Police and fire stations are another vital part of our infrastructure that we especially need during Black Sky events. The New York City Central Park Precinct installed a 200 kW fuel cell in their station prior to the huge blackout that hit the Northeast in 2003. When it hit, their station was the only building in the area to retain power[407]. Public institutions like the police and fire departments operate to serve the community, however, if they are unable to sustain themselves during an outage, how can they help others in a time of increased need? Equipping these institutions with fuel cells will allow them to operate when the grid is down to help with power restoration and assist those affected by the outage. The fuel cell also "saved the police station $200,000 in capital costs by avoiding the installation of power line extensions"[408] and was originally installed at the station to save on electrical costs.

404 Ibid

405 Ibid

406 Ibid

407 Ibid

408 Ibid

Able to supply environmentally responsible emergency power to data centers, hospitals, and critical services facilities, fuel cells have begun replacing diesel generators within microgrids. Larger scale infrastructure, from 2 MW to 15 MW is being supported with fuel cell microgrids and continuous power grid support plants at existing Northeast operating installations with FuelCell Energy molten carbonate fuel cells. Under Black Sky events, larger electrical infrastructure loads need to continue to be operational for up to weeks without the larger utility electrical grid. These critical services, which include the full operating loads at regional medical centers, trauma center hospitals, critical communications centers and other life/safety facilities can be served by larger microgrids with fuel cells, including blackstart capability for reliability. With forward looking disaster planning, fuel cell-centric microgrids are a solution to enable a direct pathway to disaster recovery even in an extreme duration event.

Fuel cells are self-starting. In this respect, they are blackstart capable. Different types of fuel cells are compatible with a variation of fuels and mixes of fuels. For instance, FuelCell Energy molten carbonate fuel cell can operate directly on renewable biogas, blends of renewable biogas and pipeline gas, and blends of hydrogen and biogas/pipeline gas. Solid oxide fuel cells can operate on hydrogen while also being capable of operating on blends of hydrogen and pipeline gas. This variety of fuel compatibility provides the emergency and disaster response planners the choices they need to balance the emergency mission needs, diversity of energy supply and the low environmental impact goals such that Black Sky disaster event is mitigated. In addition to the ability to use hydrogen gas directly, GenCell has developed a technology to extract hydrogen from Anhydrous Ammonia.

Liquid ammonia (NH3) is an excellent source of hydrogen, "containing twice as much hydrogen as liquid hydrogen by volume"[409]. Able to be stored in large tanks at room temperature, it is safer than propane and as safe as gasoline. Each year, "more than 200 million tons of ammonia are produced and distributed globally via pipelines, tankers & trucks"[410] making ammonia readily available and cheap.

409 Vinson and Elkins; Hydrogen Applications and Usage, https://www.velaw.com/insights/hydrogen-applications-and-usage/

410 GenCell; Ammonia Fuel, https://www.gencellenergy.com/gencell-technology/ammonia-fuel/

How does liquid ammonia compare to hydrogen when it comes to reusability and self-sustainability? The standard electrolysis process can produce hydrogen from water. Ammonia electrolysis is also now possible with zero-emissions using sustainable energy sources and the "process produces around 20% ammonia and 80% hydrogen"[411]. "The raw energy density of liquid ammonia is 11.5 MJ/L, higher than the 8.5 MJ/L for liquid hydrogen"[412]. Utilizing a combination of hydrogen and ammonia fuel cells with a supplemental ammonia tank and electrolysis system could prove very effective on a MW scale.

Similar to diesel, ammonia is a widely accessible liquid fuel with a developed transportation and logistical infrastructure. Superior to diesel, ammonia is "significantly less expensive than diesel, does not pollute the environment and is less likely to be stolen and resold as its use is relatively specialized"[413].

GenCell's FOX fuel cell is an ammonia power solution that provides "emission-free, reliable energy for telecoms, schools, water purification and more"[414]. Speaking to its efficiency: "a single 12-ton tank of ammonia can fuel a base station 24 hours a day, up to 365 days a year (depending on the power rating)."[415] . This makes the GenCell FOX solution ideal for use in remote locations that lack grid access or where the grid is unstable and backup power is frequently required, and a viable blackstart asset.

- **Recommendation:** Fuel cells with large ammonia fuel tanks should be prioritized as a black sky emergency power asset for critical applications.

411 The Chemical Engineer; Amanda Doyle, Lowering Emissions from Ammonia Production with Electrolysis, https://www.thechemicalengineer.com/news/lowering-emissions-from-ammonia-production-with-electrolysis/#:~:text=Using%20electricity%20derived%20from%20fossil,%25%20ammonia%20and%2080%25%20hydrogen.

412 Frontiers in Energy Research; Lan and Tao, Ammonia as a Suitable Fuel for Fuel Cells, https://www.frontiersin.org/articles/10.3389/fenrg.2014.00035/full

413 GenCell; Ammonia Fuel, https://www.gencellenergy.com/gencell-technology/ammonia-fuel/

414 GenCell; GenCell Fox Off-grid Power Solution, https://www.gencellenergy.com/our-products/gencell-a5/

415 GenCell; Ammonia Fuel, https://www.gencellenergy.com/gencell-technology/ammonia-fuel/

VI | HUMAN CONTINUITY

A. Background

As has been indicated in other sections of this handbook, blackstart capability alone, will not create societal resilience in and of itself, particularly where blackstart becomes necessary as the result of malicious action by others. A black sky event will have cascading effects across all energy dependent sectors, each of which will be managing the resulting crisis across their infrastructure and customers.

These cascading effects are likely to amplify any level of panic induced by the event. And the requirement to prioritize energy restoration to critical life-line facilities, .e.g. Hospitals, will be significant. Without careful pre-planning, the demand created by the wider impacts and prioritization of needs could easily overwhelm energy sector crisis management and response capabilities, which are likely to already be over-stretched trying to manage a black start restoration.

Even more importantly, it may not be possible to define in advance what the whole-of-grid restoration plan will look like. Restoration plans are likely to be incident specific, based on the nature and location of the disruption and the relative proximity to black start nodes to that incident.

The consequence of this is that grid restoration paths are likely to be unpredictable and so managing restoration to energy dependence sectors will need to be dynamically managed based on available resources and prevailing circumstances.

Whilst it is possible to devise generic crisis and response plans both within the energy sector and across dependent sectors, their effectiveness is likely to be limited and focused on managing the resulting consequences within their own sectors.

The infrastructures sustaining our civilization have evolved to become an efficient, increasingly complex and interdependent global system, with energy at the root.

Yet without embedded, system-wide resilience, growing complexity and interdependency means growing fragility, with serious implications for our future.

A broader all-sector capability is required and the Human Continuity Project is EIS Council's response to that broader need.

B. The wider problem: systemic, global infrastructure fragility and expanding threats

The infrastructures sustaining our civilization have evolved to become an efficient, increasingly complex and interdependent global system, with energy at the root. Yet without embedded, system-wide resilience, growing complexity and interdependency means growing fragility, with serious implications for our future.

Our world has been gripped by successive waves of globe-spanning supply chain breakdowns; by a pandemic that radically altered daily lives in every nation; by shockwaves felt around the globe from war, returning to the European continent and disrupting global trade in the Red Sea. All of this is taking place amid expanding planet-wide climate disasters, geopolitical brinksmanship, and periodic food crises with especially tragic consequences for many developing nations and disadvantaged populations.

The accelerating pace of these global crises has increased concerns over emerging Black Sky hazards, and the growing risk of unparalleled societal catastrophe.

C. The Opportunity: Investor-driven, collaborative global resilience

Based on more than a decade of systems engineering-framed multi-sector, internationally-informed research and a handful of critical, resilience-adapted tools and capabilities, leading corporations, universities, research groups and government agencies are assembling one of history's widest and most ambitious collaborations to address this civilization-scale risk.

The mission is to develop cost-effective societal, business and government continuity tools and recommendations which will enable systemic preparedness and recovery from a full range of hazards, including global-impact catastrophic events.

The mission is to develop cost-effective societal, business and government continuity tools and recommendations which will enable systemic preparedness and recovery from a full range of hazards, including global-impact catastrophic events.

1. All hazard societal – and business – continuity

Resilience on this scale represents a growing opportunity for robust, assured business continuity. At a time of exceptional growth in risk, in today's uncertain, fast-changing environment this can be a vital asset for infrastructure corporations, for their customers, and for the investors that stand behind them. Enterprises that are better able to withstand the shocks to come, including Black Sky events, will be better able to sustain their operations and avoid the cataclysmic failures that will befall less prepared businesses.

2. Regulatory policy considerations

In regulated infrastructure sectors, the Human Continuity Project's comprehensive focus on resilience will also include a thorough, analytical assessment of the effects of regulatory policies, and their impact in incentivizing – or disincentivizing – the resilience and reliability of regulated industries.

3. All-sector focused, advisor-informed

In building all-hazard resilience plans, capabilities and recommendations, project teams will work closely with advisors from all affected sectors to ensure their work products are thoroughly informed by practical, commercial realities; by changing, regionally varying regulatory landscapes; and by evolving technical considerations. This will include developing resilience criteria that can inform businesses and their suppliers, investment fund managers and others to help assess the resilience of companies against extreme hazards, in ways that go above and beyond the metrics and risk assessment analytics that currently exist.

Today, some of the most respected universities, research institutions and corporations worldwide are collaborating in the Human Continuity Project® to develop these capabilities. Infrastructure businesses, national, regional and international organizations, government agencies and other key stakeholders are partnering as advisors to help ensure the project is focused – both on practical considerations and on urgent societal priorities and social needs.

Committed, caring people and organizations. Collaborating; for all of humankind.

D. The Plan

Our society's complex infrastructure network is made up of the interconnected services of countless organizations. Yet each element of this interdependent system is managed as a relatively small, separate entity. One of the challenges in building systemic resilience for our society is that the key decision makers – the owners and operators of these organizations – have no visibility into the full system. Understandably, their most common response to any Black Sky resilience question is often simply, "But what can we do? Where can we begin?"

The core and framework of the Human Continuity Project is designed to address precisely this issue.

1. System Engineering

Adapting the methodology pioneered by the aerospace and high-tech industries to build and secure complex, high value instruments.

The aerospace industry faces a similar problem when beginning development of any new, innovative instrument. On-orbit instruments in particular, require smoothly integrated efforts of hundreds or thousands of companies and suppliers, most of whom have never worked together, each bringing its own, unique skill set to the project. The key to making this work, developed and continually refined over the last 50 years, is system engineering.

In this light, the complex infrastructures that underpin our industrial world, sustaining our society, can be viewed as a complex system subject to the same system engineering process.

By applying this methodology, first locally, then regionally and ultimately globally, we can begin an evolving process to secure the most important "instrument" ever developed: the global infrastructure network that sustains our economies, our lives, and our hopes for the future of humankind.

2. Project Structure

The Human Continuity Project® will begin with development of two major tasks in parallel.

a. The Research and Development Task is designed to develop the overall system engineering framework that will structure the work of the project.

This will include definition of the interdependent services provided by each of the interdependent infrastructure sectors and supply chains; definition of key sector interfaces; development of high-level functional models for each sector; and advanced development of the GINOM® Multi-corporate Operating System (MOS) Simulation which – using these defined services, interfaces and functional models – will enable functional integration of participating corporate simulations or operating systems. In addition, given their importance in incentivizing / disincentivizing sector by sector Black Sky resilience capabilities, this task also includes assessment of key policy requirements affecting each sector.

b. In parallel, the Initial Implementation Task is designed to expand progress in implementing Black Sky resilience recommendations where they are already available from earlier work.

This includes an expansion of the U.S. Eastern Interconnection emergency communication Energy Sector Web™ pilot to other regional locations; development of early implementation Black Sky Emergency Power pilots (i.e., adapting or supplementing existing facility emergency power systems to optimize capability for long duration outages); and building on this handbook, developing comparable Black Sky-class blackstart recommendations for other regions. Just as in the Research and Development Task, this Task also includes a key policy element, with a focus on engaging with government agencies to enable and encourage policy adaptations associated with each of these early implementation areas.

And finally, given the scale and criticality of the Project's societal continuity mission, this task will also include prioritized community engagement, as a mechanism to create public awareness of the opportunity this effort represents for our civilization.

9

CHAPTER NINE

CONCLUSION: FINDINGS, RECOMMENDATIONS, AND THE PATH AHEAD

A core mission of the Electric Infrastructure Security Council is to help modern societies strengthen their resilience against Black Sky events. To that end, a core activity is to ensure that a robust and nimble blackstart capability is available to re-energize the electric grid should electric blackouts occur - no matter the cause. As has been outlined in this Handbook, that capability is under increasing pressure. The interdependence and electrification of critical infrastructure

and core societal functions continues to grow. Moreover, as the imperative to fight climate change creates challenges for ensuring blackstart resource adequacy, it is all the more crucial for industry, government, and regulators to partner together to ensure that blackstart capabilities remain effective and survivable against all hazards.

This handbook emerges from comprehensive analysis and recommendations provided by experts throughout the energy sector, and by government policymakers, regulators, academia, and other non-government organizations.

Brief restatement: key findings and options for consensus-building on blackstart initiatives

1. **FINDING:** SOCIETAL CONTINUITY IS INCREASINGLY DEPENDENT ON RESILIENT BLACKSTART CAPABILITIES

RECOMMENDATIONS

- Build on ongoing efforts to map the interdependencies between electricity-dependent infrastructure and adjust power restoration plans accordingly.
- Develop more robust consultative mechanisms and organizational arrangements to help TOPs shape and refine their restoration plans.

2. **FINDING:** BLACKSTART RESILIENCE PLANNING MUST INCLUDE ALL PHASES OF THE RESTORATION PROCESS

RECOMMENDATIONS

- Address the resilience challenges (and solutions) for blackstart restoration across the entire process.
- Be prepared for adversaries to target emergency communications as a prime attack vector across all restoration stages.
- Draw on lessons learned from past severe events.

3. **FINDING:** GRID OPERATORS SHOULD BE READY TO CONDUCT BLACKSTART IN THE MIDST OF SOCIETAL CHAOS AND INTENSE POLITICAL PRESSURES

RECOMMENDATIONS

- Accelerate exercises and planning for Grid Security Emergencies.
- Scale up planning for law enforcement support.
- Build on the Existing Playbook System to Help Counter Enemy Information Operations.

4. **FINDING:** DANGERS TO BLACKSTART RESOURCE ADEQUACY ARE GROWING

RECOMMENDATION

- Leverage emerging best practices in state legislation and regulations that support grid reliability and resilience.

5. **FINDING:** IMMEDIATE INITIATIVES ARE REQUIRED TO ENSURE THE OPERATING RELIABILITY OF BLACKSTART IN AN IBR-HEAVY GRID

RECOMMENDATIONS

- Take stronger measures to remedy the performance failures of IBR.
- Immediately address the danger of adversary mis-operation of IBRs.

6. **FINDING:** OVER THE LONGER TERM, IBRs COULD OFFER UNIQUE BENEFITS TO BLACKSTART RESILIENCE – BUT ONLY IF THEIR TECHNICAL AND OPERATIONAL PROBLEMS CAN BE RESOLVED

RECOMMENDATIONS

- Build on lessons learned from PJM's initiatives on blackstart and IBRs.
- Develop multi-island restoration strategies that leverage accelerating hybrid plant deployments.
- Rethink the tradeoffs that CIP standards strike between the need to ensure resource adequacy, versus the imperative for generation survivability.

7. **FINDING:** GRID FOLLOWING AND GRID FORMING INVERTERS POSE DISTINCT PROBLEMS – AND DIFFERENT POTENTIAL ADVANTAGES – FOR BLACKSTART RESILIENCE

RECOMMENDATIONS

- Act now to ensure battery energy storage systems (BESS) and hybrid plants (which combine IBR generation and storage) have built-in capabilities to support restoration.
- Incorporate blackstart capabilities in the development of advanced IBR performance standards.

8. **FINDING:** MEASURES TO SECURE THE FLOW OF FUEL FOR GAS-FIRED GENERATORS MUST SCALE WITH INTENSIFYING THREATS

RECOMMENDATIONS

- Build gas-electric information sharing protocols and communications capabilities to support restoration.
- Prioritize resilience measures for secondary fuels.
- Begin developing mandatory reliability and infrastructure protection standards for natural gas infrastructure before a Black Sky event makes them inevitable – and too late.
- Natural gas-fed Blackstart and Next-start units in the U.S. should adopt a 2-layer strategy in protecting their fuel sources.

9. **FINDING:** NEW HYDROPOWER, GEOTHERMAL AND FUEL CELL INITIATIVES CAN HELP MEET EMERGING BLACKSTART REQUIREMENTS

RECOMMENDATIONS

- Establish an interagency, public-private task force to identify promising opportunities for small hydro blackstart generation. FERC and DOE should partner with industry to create specially focused initiatives to leverage these opportunities on a nationwide basis. Equivalent efforts should be launched by other nations as well.
- Existing and planned geothermal plants of sufficient size and in viable locations should be included as resources in blackstart planning.
- As Fuel Cell arrays become larger, they should be considered a blackstart asset, first at microgrid or distribution grid level, and eventually for the BES.

10. **FINDING:** EXISTING NUCLEAR POWER PRESENT CRITICAL BLACK SKY CHALLENGES, BUT COULD BECOME FUTURE BLACK SKY BLACKSTART ASSETS

RECOMMENDATIONS

- The U.S. government should review the licensing, regulations, and any technical barriers currently preventing the existing nuclear fleet from becoming Black Sky assets through more flexible operation and ensure that SMRs are designed to enable plant island mode and blackstart operations.
- The nuclear industry trade groups, Nuclear Energy Institute (NEI) and the Institute of Nuclear Power Plants Operations (INPO) should review this document for actions to improve the resiliency of U.S. nuclear power plants. This action could be in the form of a joint task force to develop the Islanding Mode and Blackstart capability at the nuclear generation units to support BES grid restoration.

The Path Ahead

As power grids transition today toward increasing dependence on distributed, green generation, we have begun seeing a steady growth in grid reliability and resilience issues, which has raised significant concern. The findings summarized in this Handbook represent comprehensive, consensus guidelines to address this issue, provided by a remarkably wide array of energy, regulatory and security subject matter experts.

As nations and as members of the global community, it has become essential that we consider and implement these findings, and move quickly to address these concerns. If we fail to do so, in addition to expanding risks to grid continuity and restoration, we also risk fostering growing resistance to the grid's green transition.

In summary, the secure path to a successful, green transition is one that is also focused on guiding that transition to ensure it enhances, rather than degrades, grid reliability and resilience.

The research and guidelines offered in this Handbook are offered in the hope they will help decision makers proceed down that path: providing a basis for serious, diligent consideration and decisive action toward a greener and robust energy future for our world.

Made in United States
Orlando, FL
04 November 2026

90138308R00205